Mike Holt's Workbook To

Understanding the National Electrical Code Volume 1

Based on the 2005 *NEC*©

www.*NEC*code.com
1.888.*NEC*®code

Mike Holt Enterprises, Inc.
1.888.NEC.CODE • NECcode.com • Info@NECcode.com

NOTICE TO THE READER

Publisher does not warrant or guarantee any of the products described herein or perform any independent analysis in connection with any of the product information contained herein. Publisher does not assume, and expressly disclaims, any obligation to obtain and include information other than that provided to it by the manufacturer.

The reader is expressly warned to consider and adopt all safety precautions that might be indicated by the activities herein and to avoid all potential hazards. By following the instructions contained herein, the reader willingly assumes all risks in connection with such instructions.

The publisher makes no representation or warranties of any kind, including but not limited to, the warranties of fitness for particular purpose or merchantability, nor are any such representations implied with respect to the material set forth herein, and the publisher takes no responsibility with respect to such material. The publisher shall not be liable for any special, consequential, or exemplary damages resulting, in whole or part, from the reader's use of, or reliance upon, this material.

Mike Holt's Workbook To Accompany
Understanding the *National Electrical Code*, Volume 1
5th Edition
Graphic Illustrations: Mike Culbreath
Cover Design: Tracy Jette
Layout Design and Typesetting: Cathleen Kwas

COPYRIGHT © 2005 Charles Michael Holt Sr.

ISBN: 1-932685-18-9

For more information, call 1-888-NEC CODE, or email info@MikeHolt.com.

All rights reserved. No part of this work covered by the copyright hereon may be reproduced or used in any form or by any means graphic, electronic, or mechanical, including photocopying, recording, taping, or information storage and retrieval systems without the written permission of the publisher. You can request permission to use material from this text, phone 1.888.NEC.CODE, Sales@NECcode.com, or www.NECcode.com.

NEC, *NFPA*, and *National Electrical Code* are registered trademarks of the National Fire Protection Association.

 This logo is a registered trademark of Mike Holt Enterprises, Inc.

I dedicate this book to the
Lord Jesus Christ,
my mentor and teacher.

To request examination copies of this or other
Mike Holt Publications, call:
Phone: 1-888-NEC CODE • Fax: 1-954-720-7944

or email:
info@MikeHolt.com

Mike Holt Online
www.NECcode.com
Sales@NECcode.com

You can download a sample PDF of all our publications by visiting www.NECcode.com

Introduction

This *Workbook to Accompany the Understanding the NEC, Volume 1* (2005 Edition) contains over 1,200 *NEC* practice questions that are in *Code* order. A final exam with 100 *NEC* questions that are in random order and discussed in the *2005 Understanding the NEC, Volume 1* textbook is also included. These questions are designed to test your knowledge and comprehension of the material covered in the textbook.

How To Use This Book

This workbook is intended to be used with the *Understanding the NEC, Volume 1* textbook and the *2005 National Electrical Code*.

Workbook Errors and Corrections

Humans develop the text and layout of this workbook, and since currently none of us is perfect, there may be a few errors. This could occur because the *NEC* is dramatically changed each *Code* cycle; new Articles are added, some deleted, some relocated, and many renumbered. In addition, this workbook must be written within a very narrow window of opportunity; after the *NEC* has been published (September), yet before it's enforceable (January).

You can be sure we work a tremendous number of hours and use all of our available resources to produce the finest product with the fewest errors. We take great care in researching the *Code* requirements to ensure this workbook is correct. If you feel there's an error of any type in this workbook (typo, grammar, or technical), no matter how insignificant, please let us know.

Any errors found after printing are listed on our Website, so if you find an error, first check to see if it has already been corrected. Go to www.MikeHolt.com, click on the "Books" link, and then the "Corrections" link (www.MikeHolt.com/book corrections.htm).

If you do not find the error listed on the Website, contact us by E-mailing corrections@MikeHolt.com, calling 1.888.NEC.CODE (1.888.632.2633), or faxing 954.720.7944. Be sure to include the book title, page number, and any other pertinent information.

Internet

Today as never before, you can get your technical questions answered by posting them to Mike Holt's Code Forum. Just visit www.MikeHolt.com and click on the "Code Forum" link.

Different Interpretations

Some electricians, contractors, instructors, inspectors, engineers, and others enjoy the challenge of discussing the *Code* requirements, hopefully in a positive and a productive manner. This action of challenging each other is important to the process of better understanding the *NEC*'s requirements and its intended application. However, if you're going to get into an *NEC* discussion, please do not spout out what you think without having the actual *Code* in your hand. The professional way of discussing an *NEC* requirement is by referring to a specific section, rather than by talking in vague generalities.

The National Electrical Code

The *National Electrical Code (NEC)* is written for persons who understand electrical terms, theory, safety procedures, and electrical trade practices. These individuals include electricians, electrical contractors, electrical inspectors, electrical engineers, designers, and other qualified persons. The *Code* was not written to serve as an instructive or teaching manual for untrained individuals [90.1(C)].

Learning to use the *NEC* is somewhat like learning to play the game of chess; it's a great game if you enjoy mental warfare. You must first learn the names of the game pieces, how the pieces are placed on the board, and how each piece moves.

In the electrical world, this is equivalent to completing a comprehensive course on basic electrical theory, such as:

- What electricity is and how is it produced
- Dangers of electrical potential: fire, arc blast, arc fault, and electric shock
- Direct current
- Series and parallel circuits
- Electrical formulas
- Alternating current
- Induction, motors, generators, and transformers

Once you understand the fundamentals of the game of chess, you're ready to start playing the game. Unfortunately, at this point all you can do is make crude moves, because you really do not understand how all the information works together. To play chess well, you will need to learn how to use your knowledge by working on subtle strategies before you can work your way up to the more intriguing and complicated moves.

Again, back to the electrical world, this is equivalent to completing a course on the basics of electrical theory. You have the foundation upon which to build, but now you need to take it to the next level, which you can do by reading the Understanding the National Electrical Code, Volume 1 textbook, watching its companion video or DVD, and answering the *NEC* practice questions in this workbook.

Not a Game

Electrical work isn't a game, and it must be taken very seriously. Learning the basics of electricity, important terms and concepts, as well as the basic layout of the *NEC* gives you just enough knowledge to be dangerous. There are thousands of specific and unique applications of electrical installations, and the *Code* doesn't cover every one of them. To safely apply the *NEC*, you must understand the purpose of a rule and how it affects the safety aspects of the installation.

NEC Terms and Concepts

The *NEC* contains many technical terms, so it's crucial that *Code* users understand their meanings and their applications. If you do not understand a term used in a *Code* rule, it will be impossible to properly apply the *NEC* requirement. Be sure you understand that Article 100 defines the terms that apply to *two or more* Articles. For example, the term "Dwelling Unit" applies to many Articles. If you do not know what a Dwelling Unit is, how can you possibly apply the *Code* requirements for it?

In addition, many Articles have terms that are unique for that specific Article. This means that the definition of those terms is only applicable for that given Article. For example, Article 250 Grounding and Bonding has the definitions of a few terms that are only to be used within Article 250.

Small Words, Grammar, and Punctuation

It's not only the technical words that require close attention, because even the simplest of words can make a big difference to the intent of a rule. The word "or" can imply alternate choices for equipment wiring methods, while "and" can mean an additional requirement. Let's not forget about grammar and punctuation. The location of a comma "," can dramatically change the requirement of a rule.

Slang Terms or Technical Jargon

Electricians, engineers, and other trade-related professionals use slang terms or technical jargon that isn't shared by all. This makes it very difficult to communicate because not everybody understands the intent or application of those slang terms. So where possible, be sure you use the proper word, and do not use a word if you do not understand its definition and application. For example, lots of electricians use the term "pigtail" when describing the short conductor for the connection of a receptacle, switch, luminaire, or equipment. Although they may understand it, not everyone does. **Figure 1**

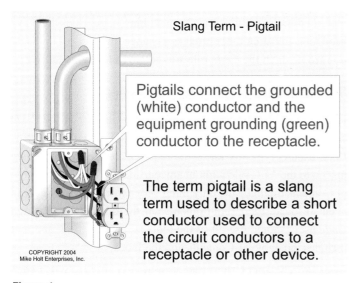

Figure 1

NEC Style and Layout

Before we get into the details of the *NEC*, we need to take a few moments to understand its style and layout. Understanding the structure and writing style of the *Code* is very important before it can be used effectively. If you think about it, how can you use something if you don't know how it works? Okay, let's get started. The *National Electrical Code* is organized into nine components.

- Table of Contents
- Chapters 1 through 9 (major categories)
- Articles 90 through 830 (individual subjects)
- Parts (divisions of an Article)
- Sections and Tables (*Code* requirements)
- Exceptions (*Code* permissions)
- Fine Print Notes (explanatory material)
- Index
- Annexes (information)

1. Table of Contents. The Table of Contents displays the layout of the Chapters, Articles, and Parts as well as the page numbers. It's an excellent resource and should be referred to periodically to observe the interrelationship of the various *NEC* components.

When attempting to locate the rules for a particular situation, knowledgeable *Code* users often go first to the Table of Contents to quickly find the specific *NEC* section that applies.

2. Chapters. There are nine Chapters, each of which is divided into Articles. The Articles fall into one of four groupings: General Requirements (Chapters 1 through 4), Specific Requirements (Chapters 5 through 7), Communications Systems (Chapter 8), and Tables (Chapter 9).

- Chapter 1 General
- Chapter 2 Wiring and Protection
- Chapter 3 Wiring Methods and Materials
- Chapter 4 Equipment for General Use
- Chapter 5 Special Occupancies
- Chapter 6 Special Equipment
- Chapter 7 Special Conditions
- Chapter 8 Communications Systems (Telephone, Data, Satellite, and Cable TV)
- Chapter 9 Tables—Conductor and Raceway Specifications

3. Articles. The *NEC* contains approximately 140 Articles, each of which covers a specific subject. For example:

- Article 110 General Requirements
- Article 250 Grounding
- Article 300 Wiring Methods
- Article 430 Motors
- Article 500 Hazardous (Classified) Locations
- Article 680 Swimming Pools, Spas, Hot Tubs, and Fountains
- Article 725 Remote-Control, Signaling, and Power-Limited Circuits
- Article 800 Communications Systems

4. Parts. Larger Articles are subdivided into Parts. For example, Article 110 has been divided into multiple parts:

- Part I. General (Sections 110.1—110.23)
- Part II. 600 Volts, Nominal, or Less (110.26 – 110.27)
- Part III. Over 600 Volts, Nominal (110.30—110.59)

Note: Because the Parts of a *Code* Article aren't included in the Section numbers, we have a tendency to forget what "Part" the *NEC* rule is relating to. For example, Table 110.34(A) contains the working space clearances for electrical equipment. If we aren't careful, we might think this table applies to all electrical installations, but Table 110.34(A) is located in Part III, which contains the requirements for Over 600 Volts, Nominal installations. The rules for working clearances for electrical equipment for systems 600V or less are contained in Table 110.26(A)(1), which is located in Part II. 600 Volts, Nominal, or Less.

5. Sections and Tables.

Sections: Each *NEC* rule is called a *Code* Section. A *Code* Section may be broken down into subsections by letters in parentheses (A), (B), etc. Numbers in parentheses (1), (2), etc., may further break down a subsection, and lower-case letters (a), (b), etc., further break the rule down to the third level. For example, the rule requiring all receptacles in a dwelling unit bathroom to be GFCI protected is contained in Section 210.8(A)(1). Section 210.8(A)(1) is located in Chapter 2, Article 210, Section 8, sub-section (A), sub-subsection (1).

Many in the industry incorrectly use the term "Article" when referring to a *Code* Section. For example, they say "Article 210.8," when they should say "Section 210.8."

Tables: Many *Code* requirements are contained within Tables, which are lists of *NEC* requirements placed in a systematic arrangement. The titles of the Tables are extremely important; they must be carefully read in order to understand the contents, applications, limitations, etc., of each Table in the *Code*. Many times notes are provided in a table; be sure to read them as well, since they are also part of the requirement. For example, Note 1 for Table 300.5 explains how to measure the cover when burying cables and raceways, and Note 5 explains what to do if solid rock is encountered.

6. Exceptions.
Exceptions are *Code* requirements that provide an alternative method to a specific requirement. There are two types of exceptions—mandatory and permissive. When a rule has several exceptions, those exceptions with mandatory requirements are listed before the permissive exceptions.

Mandatory Exception: A mandatory exception uses the words "shall" or "shall not." The word "shall" in an exception means that if you're using the exception, you're required to do it in a particular way. The term "shall not" means it isn't permitted.

Permissive Exception: A permissive exception uses words such as "is permitted," which means that it's acceptable to do it in this way.

7. Fine Print Note (FPN).
A Fine Print Note contains explanatory material intended to clarify a rule or give assistance, but it isn't a *Code* requirement.

8. Index.
The Index contained in the *NEC* is excellent and is helpful in locating a specific rule.

9. Annexes.
Annexes aren't a part of the *NEC* requirements, and are included in the *Code* for informational purposes only.

- Annex A. Product Safety Standards
- Annex B. Application Information for Ampacity Calculation
- Annex C. Conduit and Tubing Fill Tables for Conductors and Fixture Wires of the Same Size
- Annex D. Examples
- Annex E. Types of Construction
- Annex F. Cross-Reference Tables (1999, 2002, and 2005 *NEC*)
- Annex G. Administration and Enforcement

Note: Changes to the *NEC*, since the previous edition(s) are identified in the margins by a vertical line (|), but rules that have been relocated aren't identified as a change. In addition, the location from which the *Code* rule was removed has no identifier.

How to Locate a Specific Requirement

How to go about finding what you're looking for in the *Code* depends, to some degree, on your experience with the *NEC*. *Code* experts typically know the requirements so well that they just go to the *NEC* rule without any outside assistance. The Table of Contents might be the only thing very experienced *Code* users need to locate their requirement. On the other hand, average *Code* users should use all of the tools at their disposal, and that includes the Table of Contents and the Index.

Table of Contents: Let's work out a simple example: What *NEC* rule specifies the maximum number of disconnects permitted for a service? If you're an experienced *Code* user, you'll know that Article 230 applies to "Services," and because this Article is so large, it's divided up into multiple parts (actually 8 parts). With this knowledge, you can quickly go to the Table of Contents (page 70-2) and see that it lists the Service Equipment Disconnecting Means requirements in Part VI, starting at page 70-77.

Note: The number 70 precedes all page numbers because the *NEC* is standard number 70 within the collection of *NFPA* standards.

Index: If you used the Index, which lists subjects in alphabetical order, to look up the term "service disconnect," you would see that there's no listing. If you tried "disconnecting means," then "services," you would find the Index specifies that the rule is located at 230, Part VI. Because the *NEC* doesn't give a page number in the Index, you'll need to use the Table of Contents to get the page number, or flip through the *Code* to Article 230, then continue to flip until you find Part VI.

As you can see, although the index is very comprehensive, it's not that easy to use if you do not understand how the index works. But if you answer the over 1,200 *NEC* practice questions contained in this workbook, you'll become a master at finding things in the *Code* quickly.

Many people complain that the *NEC* only confuses them by taking them in circles. As you gain experience in using the *Code* and deepen your understanding of words, terms, principles, and practices, you will find the *NEC* much easier to understand and use than you originally thought.

Customizing Your *Code* Book

One way to increase your comfort level with the *Code* is to customize it to meet your needs. You can do this by highlighting and underlining important *NEC* requirements, and by attaching tabs to important pages.

Highlighting: As you answer the questions in this workbook, and read through the *Understanding the National Electrical Code, Volume 1* textbook, be sure you highlight those requirements in the *Code* that are most important to you. Use yellow for general interest and orange for important requirements you want to find quickly. Be sure to highlight terms in the Index and Table of Contents as you use them.

Because of the size of the 2005 *NEC*, I recommend you highlight in green the Parts of Articles that are important for your applications, particularly:

Article 230 Services
Article 250 Grounding
Article 430 Motors

Underlining: Underline or circle key words and phrases in the *NEC* with a red pen (not a lead pencil) and use a 6 in. ruler to keep lines straight and neat. This is a very handy way to make important requirements stand out. A small 6 in. ruler also comes in handy for locating specific information in the many *Code* tables.

Tabbing the *NEC*: Placing tabs on important *Code* Articles, Sections, and Tables will make it very easy to access important *NEC* requirements. However, too many tabs will defeat the purpose. You can order a custom set of *Code* tabs, designed by Mike Holt, online at www.MikeHolt.com, or by calling us at 1.888.*NEC*.Code (1.888.632.2633).

About the Author

Mike Holt worked his way up through the electrical trade from an apprentice electrician to become one of the most recognized experts in the world as it relates to electrical power installation. He was a Journeyman Electrician, Master Electrician, and Electrical Contractor. Mike came from the real world, and he has a unique understanding of how the *NEC* relates to electrical installations from a practical standpoint. You will find his writing style to be simple, nontechnical, and practical.

Did you know that Mike didn't finish high school? So if you struggled in high school or if you didn't finish it at all, don't let this get you down, you're in good company. As a matter of fact, Mike Culbreath, Master Electrician, who produces the finest electrical graphics in the history of the electrical industry, didn't finish high school either! So two high school dropouts produce the text and graphics in Mike Holt's textbooks! However, realizing that success depends on one's continuing pursuit of education, Mike immediately attained his GED (as did Mike Culbreath) and ultimately attended the University of Miami's Graduate School for a Master's degree in Business Administration (MBA).

Mike Holt resides in Central Florida, is the father of seven children, and has many outside interests and activities. He is a former National Barefoot Waterskiing Champion (1988 and 1999), who set five barefoot water-ski records, and he continues to train year-round at a national competition level [www.barefootcentral.com].

Mike enjoys motocross racing, but at the age of 52 decided to retire from that activity (way too many broken bones, concussions, collapsed lung, etc., but what a rush). Mike also enjoys snow skiing and spending time with his family. What sets Mike apart from some is his commitment to living a balanced lifestyle; he places God first, then family, career, and self.

Questions for Article 90—Introduction

Article 90 Introduction to the *National Electrical Code*

Article Overview

Many *NEC* violations and misunderstandings wouldn't occur if people doing the work simply understood Article 90. For example, many people see *Code* requirements as performance standards. In fact, *NEC* requirements are the bare minimum for safety. This is exactly the stance electrical inspectors, insurance companies, and courts will take when making a decision regarding electrical design or installation.

Article 90 opens by saying the *NEC* isn't intended as a design specification or instruction manual. The *National Electrical Code* has one purpose only. That is "the practical safeguarding of persons and property from hazards arising from the use of electricity."

Article 90 then describes the scope and arrangement of the *Code*. A person who says, "I can't find anything in the *Code*," is really saying, "I never took the time to review Article 90." The balance of Article 90 provides the reader with information essential to understanding those items you do find in the *NEC*.

Typically, electrical work requires you to understand the first four Chapters of the *NEC*, plus have a working knowledge of the Chapter 9 tables. Chapters 5, 6, 7, and 8 make up a large portion of the *NEC*, but they apply to special situations. They build on, and extend, what you must know in the first four chapters. That knowledge begins with Chapter 1 in the *Understanding the NEC, Volume 1* textbook.

Questions

1. The *NEC* is _____.

    (a) intended to be a design manual
    (b) meant to be used as an instruction guide for untrained persons
    (c) for the practical safeguarding of persons and property
    (d) published by the Bureau of Standards

2. Compliance with the provisions of the *Code* will result in _____.

    (a) good electrical service
    (b) an efficient electrical system
    (c) an electrical system essentially free from hazard
    (d) all of these

3. •The *Code* contains provisions considered necessary for safety, which will not necessarily result in _____.

    (a) efficient use
    (b) convenience
    (c) good service or future expansion of electrical use
    (d) all of these

(• Indicates that 75% or fewer exam takers get the question correct)

4. Hazards often occur because of _____.

   (a) overloading of wiring systems by methods or usage not in conformity with this *Code*
   (b) initial wiring not providing for increases in the use of electricity
   (c) a and b
   (d) none of these

5. The following systems must be installed in accordance with the *NEC*:

   (a) signaling   (b) communications   (c) power and lighting   (d) all of these

6. •The *Code* applies to the installation of _____.

   (a) electrical conductors and equipment within or on public and private buildings
   (b) outside conductors and equipment on the premises
   (c) optical fiber cable
   (d) all of these

7. The *Code* does not cover installations in ships, watercraft, railway rolling stock, aircraft, or automotive vehicles.

   (a) True   (b) False

8. The *Code* covers underground installations in mines and self-propelled mobile surface mining machinery and its attendant electrical trailing cable.

   (a) True   (b) False

9. Installations of communications equipment that are under the exclusive control of communications utilities, and located outdoors or in building spaces used exclusively for such installations _____ covered by the *Code*.

   (a) are   (b) are sometimes   (c) are not   (d) might be

10. The *Code* covers all of the following electrical installations except _____.

    (a) floating buildings
    (b) in or on private and public buildings
    (c) industrial substations
    (d) electrical generation installations on property owned or leased by the electric utility company

11. Service laterals installed by an electrical contractor must be installed in accordance with the *NEC*.

    (a) True   (b) False

12. Utilities include entities that install, operate, and maintain _____.

    (a) communications systems (telephone, CATV, Internet, satellite, or data services)
    (b) electric supply systems (generation, transmission, or distribution systems)
    (c) Local Area Network wiring on premises
    (d) a or b

13. Utilities may be subject to compliance with codes and standards covering their regulated activities as adopted under governmental law or regulation.

    (a) True   (b) False

(• Indicates that 75% or fewer exam takers get the question correct)

14. Utilities include entities that are designated or recognized by governmental law or regulation by public service/utility commissions.

    (a) True   (b) False

15. Communications wiring such as telephone, antenna, and CATV wiring within a building is not required to comply with the installation requirements of Chapters 1 through 7, except where it is specifically referenced therein.

    (a) True   (b) False

16. The requirements in "Annexes" must be complied with.

    (a) True   (b) False

17. The authority having jurisdiction (AHJ) has the responsibility _____.

    (a) for making interpretations of the rules of the *Code*
    (b) for deciding upon the approval of equipment and materials
    (c) for waiving specific requirements in the *Code* and allowing alternate methods and material if safety is maintained
    (d) all of these

18. In order for equipment and materials to receive approval _____.

    (a) the equipment and material must always be listed by UL
    (b) the authority having jurisdiction must decide on approval
    (c) the authority having jurisdiction can never approve non-listed items
    (d) the local electrical distributor must decide on approval

19. A *Code* rule may be waived or alternative methods of installation approved that may be contrary to the *NEC*, if the authority having jurisdiction gives verbal or written consent.

    (a) True   (b) False

20. In the event the *Code* requires new products, constructions, or materials that are not yet available at the time a new edition is adopted, the _____ may permit the use of the products, constructions, or materials that comply with the most recent previous edition of this *Code* adopted by the jurisdiction.

    (a) architect   (b) master electrician
    (c) authority having jurisdiction   (d) supply house

21. Explanatory material, such as references to other standards, references to related sections of the *NEC*, or information related to a *Code* rule, are included in the form of Fine Print Notes (FPNs).

    (a) True   (b) False

22. Equipment listed by a qualified electrical testing laboratory is not required to have the factory-installed _____ wiring inspected at the time of installation except to detect alterations or damage.

    (a) external   (b) associated   (c) internal   (d) all of these

(• Indicates that 75% or fewer exam takers get the question correct)

Questions for Chapter 1—General

Article 100—Definitions

Article Overview

Have you ever had a conversation with someone, only to discover what you said and what he/she heard were completely different? This happens when one or more of the people in a conversation do not understand the definitions of the words being used, and that's why the definitions of key terms are located right up in the front of the *NEC,* in Article 100.

If we can all agree on important definitions, then we speak the same language and avoid misunderstandings. Because the *Code* exists to protect people and property, we can agree it's very important to know the definitions presented in Article 100.

Now, here are a couple of things you may not know about Article 100:

- Article 100 contains the definitions of many, but not all, of the terms used throughout the *NEC*. In general, only those terms used in two or more articles are defined in Article 100.
- Part I of Article 100 contains the definitions of terms used throughout the *Code*.
- Part II of Article 100 contains only terms that apply to systems that operate at over 600V.

How can you possibly learn all these definitions? There seem to be so many. Here are a few tips:

- Break the task down. Study a few words at a time, rather than trying to learn them all at one sitting.
- Review the graphics in the *Understanding the NEC, Volume 1* textbook. These will help you see how a term is applied.
- Relate them to your work. As you read a word, think of how it applies to the work you're doing. This will provide a natural reinforcement of the learning process.

Questions

1. Admitting close approach, not guarded by locked doors, elevation, or other effective means, is commonly referred to as _____.

   (a) accessible (equipment)
   (b) accessible (wiring methods)
   (c) accessible, readily
   (d) all of these

2. Capable of being reached quickly for operation, renewal, or inspections without resorting to portable ladders and such is known as _____.

   (a) accessible (equipment)
   (b) accessible (wiring methods)
   (c) accessible, readily
   (d) all of these

3. Capable of being removed or exposed without damaging the building structure or finish, or not permanently closed in by the structure or finish of the building defines _____.

   (a) accessible (equipment)
   (b) accessible (wiring methods)
   (c) accessible, readily
   (d) all of these

(• Indicates that 75% or fewer exam takers get the question correct)

4. A junction box located above a suspended ceiling having removable panels is considered to be _____.

   (a) concealed　　(b) accessible　　(c) readily accessible　　(d) recessed

5. Acceptable to the authority having jurisdiction means _____.

   (a) identified　　(b) listed　　(c) approved　　(d) labeled

6. •A device that, by insertion in a receptacle, establishes a connection between the conductors of the attached flexible cord and the conductors connected permanently to the receptacle is called a(n) _____.

   (a) attachment plug　　(b) plug cap　　(c) plug　　(d) any of these

7. Where no statutory requirement exists, the authority having jurisdiction could be a property owner or his/her agent, such as an architect or engineer.

   (a) True　　(b) False

8. The connection between the grounded neutral conductor and the equipment grounding conductor at the service is accomplished by installing a(n) _____ jumper.

   (a) main bonding　　(b) bonding　　(c) equipment bonding　　(d) circuit bonding

9. The connection between the grounded circuit conductor and the equipment grounding conductor at a separately derived system is the _____.

   (a) main bonding jumper　　(b) system bonding jumper
   (c) circuit bonding jumper　　(d) equipment bonding jumper

10. A branch circuit that supplies only one utilization equipment is a(n) _____ branch circuit.

    (a) individual　　(b) general-purpose　　(c) isolated　　(d) special-purpose

11. •For a circuit to be considered a multiwire branch circuit, it must have _____.

    (a) two or more ungrounded conductors with a voltage potential between them
    (b) a grounded neutral conductor having equal voltage potential between it and each ungrounded conductor of the circuit
    (c) a grounded neutral conductor connected to the grounded neutral terminal of the system
    (d) all of these

12. The *Code* defines a(n) _____ as a structure that stands alone or that is cut off from adjoining structures by firewalls, with all openings therein protected by approved fire doors.

    (a) unit　　(b) apartment　　(c) building　　(d) utility

13. A circuit breaker is a device designed to _____ a circuit by nonautomatic means and to open the circuit automatically on a predetermined overcurrent without damage to itself when properly applied within its rating.

    (a) blow　　(b) disconnect　　(c) connect　　(d) open and close

14. _____ is a qualifying term indicating that there is a purposely-introduced delay in the tripping action of the circuit breaker, which decreases as the magnitude of the current increases.

    (a) Adverse-time　　(b) Inverse-time　　(c) Time delay　　(d) Timed unit

(• Indicates that 75% or fewer exam takers get the question correct)

15. The localization of an overcurrent condition to restrict outages to the circuit or equipment affected, accomplished by the choice of overcurrent-protective devices is called _____.

   (a) overcurrent protection   (b) interrupting capacity   (c) selective coordination   (d) overload protection

16. NM cable is considered _____ if rendered inaccessible by the structure or finish of the building.

   (a) inaccessible   (b) concealed   (c) hidden   (d) enclosed

17. A separate portion of a conduit or tubing system that provides access through a removable cover(s) to the interior of the system at a junction of two or more sections of the system, or at a terminal point of the system, is defined as a(n) _____.

   (a) junction box   (b) accessible raceway   (c) conduit body   (d) pressure connector

18. A solderless pressure connector is a device that _____ between two or more conductors or between one or more conductors and a terminal by means of mechanical pressure and without the use of solder.

   (a) provides access   (b) protects the wiring   (c) is never needed   (d) establishes a connection

19. A load is considered to be continuous if the maximum current is expected to continue for _____ or more.

   (a) ½ hour   (b) 1 hour   (c) 2 hours   (d) 3 hours

20. A _____ is a device or group of devices that serves to govern in some predetermined manner the electric power delivered to the apparatus to which it is connected.

   (a) relay   (b) breaker   (c) transformer   (d) controller

21. A component of an electrical system that is intended to carry or control but not utilize electric energy is a(n) _____.

   (a) raceway   (b) fitting   (c) device   (d) enclosure

22. Which of the following does the *Code* recognize as a device?

   (a) Switch   (b) Light bulb   (c) Transformer   (d) Motor

23. A(n) _____ is a device, group of devices, or other means by which the conductors of a circuit can be disconnected from their source of supply.

   (a) feeder   (b) enclosure   (c) disconnecting means   (d) conductor interrupter

24. Continuous duty is defined as _____.

   (a) when the load is expected to continue for three hours or more
   (b) operation at a substantially constant load for an indefinite length of time
   (c) operation at loads and for intervals of time, both of which may be subject to wide variations
   (d) operation at which the load may be subject to maximum current for six hours or more

25. A _____ is a single unit that provides independent living facilities for persons, including permanent provisions for living, sleeping, cooking, and sanitation.

   (a) two-family dwelling   (b) one-family dwelling   (c) dwelling unit   (d) multifamily dwelling

26. The *NEC* term to define wiring methods that are not concealed is _____.

   (a) open   (b) uncovered   (c) exposed   (d) bare

(• Indicates that 75% or fewer exam takers get the question correct)

27. The *Code* defines a _____ as, "all circuit conductors between the service equipment, the source of a separately derived system, or other power supply source and the final branch-circuit overcurrent device."

   (a) feeder   (b) branch circuit   (c) service   (d) all of these

28. A _____ is a building or portion of a building in which one or more self-propelled vehicles can be kept for use, sale, storage, rental, repair, exhibition, or demonstration purposes.

   (a) garage   (b) residential garage   (c) service garage   (d) commercial garage

29. Connected to earth or to some conducting body that serves in place of the earth is called _____.

   (a) grounding   (b) bonded   (c) grounded   (d) all of these

30. A system or circuit conductor that is intentionally grounded is a(n) _____.

   (a) grounding conductor   (b) unidentified conductor   (c) grounded neutral conductor   (d) none of these

31. _____ is defined as intentionally connected to earth through a ground connection or connections of sufficiently low impedance and having sufficient current-carrying capacity, to prevent the build up of voltages that may result in undue hazards to connected equipment or to persons.

   (a) Effectively grounded   (b) A proper wiring system
   (c) A lighting rod   (d) A grounded neutral conductor

32. The intentional electrical connection of one system terminal to ground without the insertion of any resistor or impedance device is _____.

   (a) grounded   (b) solidly grounded   (c) effectively grounded   (d) grounding conductor

33. A "Class A" GFCI protection device is designed to de-energize the circuit when the ground-fault current is approximately _____.

   (a) 4 mA   (b) 5 mA   (c) 6 mA   (d) any of these

34. A device intended for the protection of personnel, that functions to de-energize a circuit or portion thereof within an established period of time when a current-to-ground exceeds the values established for a "Class A Device," is a(n) _____.

   (a) dual-element fuse   (b) inverse-time breaker   (c) ground-fault circuit interrupter   (d) safety switch

35. A device that establishes an electrical connection to the earth is the _____.

   (a) grounding electrode conductor   (b) grounding conductor
   (c) grounding electrode   (d) grounded neutral conductor

36. The grounding electrode conductor is the conductor used to connect the grounding electrode to the equipment grounding conductor and the grounded neutral conductor at _____.

   (a) the service   (b) each building or structure supplied by feeder(s)
   (c) the source of a separately derived system   (d) all of these

37. •In a grounded system, the conductor that connects the grounded conductor of a service, a feeder supplying a separate building or structure, or the source of a separately derived system to the grounding electrode is called the _____ conductor.

   (a) main grounding   (b) common main   (c) equipment grounding   (d) grounding electrode

(• Indicates that 75% or fewer exam takers get the question correct)

38. A _____ is an accommodation that combines living, sleeping, sanitary, and storage facilities.

    (a) guest room   (b) guest suite   (c) dwelling unit   (d) single family dwelling

39. A _____ is an accommodation with two or more contiguous rooms comprising a compartment, with or without doors between such rooms, that provides living, sleeping, sanitary, and storage facilities.

    (a) guest room   (b) guest suite   (c) dwelling unit   (d) single family dwelling

40. A handhole enclosure is an enclosure identified for use in underground systems, provided with an open or closed bottom, and sized to allow personnel to _____, for the purpose of installing, operating, or maintaining equipment or wiring or both.

    (a) enter and exit freely   (b) reach into but not enter   (c) have full working space   (d) examine visually

41. Recognized as suitable for the specific purpose, function, use, environment, and application is the definition of _____.

    (a) labeled   (b) identified (as applied to equipment)
    (c) listed   (d) approved

42. The highest current at rated voltage that a device is intended to interrupt under standard test conditions is the _____.

    (a) interrupting rating   (b) manufacturer's rating   (c) interrupting capacity   (d) GFCI rating

43. Equipment or materials to which a symbol or other identifying mark of a product evaluation organization that is acceptable to the authority having jurisdiction has been attached is known as _____.

    (a) listed   (b) labeled   (c) approved   (d) rated

44. An outlet intended for the direct connection of a lampholder, a luminaire, or a pendant cord terminating in a lampholder is a(n) _____.

    (a) outlet   (b) receptacle outlet   (c) lighting outlet   (d) general-purpose outlet

45. The environment of a wiring method under the eaves of a house having a roofed open porch would be considered a _____ location.

    (a) dry   (b) damp   (c) wet   (d) moist

46. A _____ location may be temporarily subject to dampness and wetness.

    (a) dry   (b) damp   (c) moist   (d) wet

47. Conduit installed underground or encased in concrete slabs that are in direct contact with the earth is considered a _____ location.

    (a) dry   (b) damp   (c) wet   (d) moist

48. The term "luminaire" includes "fixture(s)" and "lighting fixture(s)."

    (a) True   (b) False

49. A(n) _____ is a point on the wiring system at which current is taken to supply utilization equipment.

    (a) box   (b) receptacle   (c) outlet   (d) device

(• Indicates that 75% or fewer exam takers get the question correct)

50. Outline lighting may include an arrangement of _____ to outline or call attention to certain features such as the shape of a building or the decoration of a window.

    (a) incandescent lamps
    (b) electric-discharge lighting
    (c) electrically powered light sources
    (d) a, b, or c

51. Outline lighting may not include light sources such as light emitting diodes (LEDs).

    (a) True
    (b) False

52. Any current in excess of the rated current of equipment, or the ampacity of a conductor, is called _____.

    (a) trip current
    (b) faulted
    (c) overcurrent
    (d) shorted

53. An overload is the same thing as a short circuit or ground fault.

    (a) True
    (b) False

54. A single panel or group of panel units designed for assembly in the form of a single panel is called a _____.

    (a) switchboard
    (b) disconnect
    (c) panelboard
    (d) switch

55. The *Code* defines a(n) _____ as one familiar with the construction and operation of the electrical equipment and installations, and who has received safety training on the hazards involved.

    (a) inspector
    (b) master electrician
    (c) journeyman electrician
    (d) qualified person

56. NFPA 70E, *Standard for Electrical Safety in the Workplace,* provides information to help determine the electrical safety training requirements expected of a "qualified person."

    (a) True
    (b) False

57. Something constructed, protected, or treated so as to prevent rain from interfering with the successful operation of the apparatus under specified test conditions is defined as _____.

    (a) raintight
    (b) waterproof
    (c) weathertight
    (d) rainproof

58. A raintight enclosure is constructed or protected so that exposure to a beating rain will not result in the entrance of water under specified test conditions.

    (a) True
    (b) False

59. A contact device installed at an outlet for the connection of an attachment plug is known as a(n) _____.

    (a) attachment point
    (b) tap
    (c) receptacle
    (d) wall plug

60. A single receptacle is a single contact device with no other contact device on the same _____.

    (a) circuit
    (b) yoke
    (c) run
    (d) equipment

61. When one electrical circuit controls another circuit through a relay, the first circuit is called a _____.

    (a) control circuit
    (b) remote-control circuit
    (c) signal circuit
    (d) controller

(• Indicates that 75% or fewer exam takers get the question correct)

62. A(n) _____ system is a premises wiring system whose power is derived from a source of electric energy or equipment other than a service, and that has no direct electrical connection, including a solidly connected grounded circuit conductor, to supply conductors originating in another system.

    (a) separately derived   (b) classified   (c) direct   (d) emergency

63. Service conductors only originate from the service point and terminate at the service equipment (disconnect).

    (a) True   (b) False

64. Overhead-service conductors from the last pole or other aerial support to and including the splices, if any, are called _____ conductors.

    (a) service-entrance   (b) service-drop   (c) service   (d) overhead service

65. The service conductors between the terminals of the service equipment and a point usually outside the building, clear of building walls, where they are joined by tap or splice to the service drop are called _____ service entrance conductors.

    (a) underground   (b) complete   (c) overhead   (d) grounded

66. The _____ is the necessary equipment, usually consisting of a circuit breaker(s) or switch(es) and fuse(s) and their accessories, connected to the load end of service conductors to a building or other structure, or an otherwise designated area, and intended to constitute the main control and cutoff of the supply.

    (a) service equipment   (b) service
    (c) service disconnect   (d) service overcurrent protection device

67. •Underground service conductors between the street main and the first point of connection to the service entrance are known as the _____.

    (a) utility service   (b) service lateral   (c) service drop   (d) main service conductors

68. The _____ is the point of connection between the facilities of the serving utility and the premises wiring.

    (a) service entrance   (b) service point
    (c) overcurrent protection   (d) beginning of the wiring system

69. A signaling circuit is any electric circuit that energizes signaling equipment.

    (a) True   (b) False

70. A(n) _____ is intended to provide limited overcurrent protection for specific applications and utilization equipment, such as luminaires and appliances. This limited protection is in addition to the protection provided by the required branch circuit overcurrent protective device.

    (a) supplementary overcurrent protective device   (b) transient voltage surge suppressor
    (c) arc-fault circuit interrupter   (d) Class A GFCI

71. A form of general-use switch constructed so that it can be installed in device boxes or on box covers, or otherwise used in conjunction with wiring systems recognized by the *Code*, is called a _____ switch.

    (a) transfer   (b) motor-circuit   (c) general-use snap   (d) bypass isolation

(• Indicates that 75% or fewer exam takers get the question correct)

72. The voltage of a circuit is defined by the *Code* as the _____ root-mean-square (effective) difference of potential between any two conductors of the circuit.

    (a) lowest   (b) greatest   (c) average   (d) nominal

73. A value assigned to a circuit or system for the purpose of conveniently designating its voltage class such as 120/240V is called _____ voltage.

    (a) root-mean-square   (b) circuit   (c) nominal   (d) source

74. An enclosure or device constructed so that moisture will not enter the enclosure or device under specific test conditions is called _____.

    (a) watertight   (b) moistureproof   (c) waterproof   (d) rainproof

75. A(n) _____ enclosure is so constructed or protected that exposure to the weather will not interfere with successful operation.

    (a) weatherproof   (b) weathertight   (c) weather-resistant   (d) all weather

(• Indicates that 75% or fewer exam takers get the question correct)

Article 110 Requirements for Electrical Installations

Article Overview

Article 110 sets the stage for how you will implement the rest of the *NEC*. This article contains a few of the most important and yet neglected parts of the *Code*. For example:

- What do you do with unused openings in enclosures?
- What's the right working clearance for a given installation?
- How should you terminate conductors?
- What kinds of warnings, markings, and identification does a given installation require?

It's critical that you master Article 110, and that's exactly what the *Understanding the NEC, Volume 1* textbook is designed to help you do. As you read this article, remember that doing so helps build your foundation for correctly applying much of the *NEC*. In fact, the article itself is a foundation for much of the *Code*. You may need to read something several times to understand it. The time you take to do that will be well spent. The illustrations in the textbook will also help. But if you find your mind starting to wander, take a break. What matters is how well you master the material and how safe your work is—not how fast you blazed through a book.

Questions

1. In determining equipment to be installed, considerations such as the following should be evaluated:

   (a) Mechanical strength   (b) Cost   (c) Arcing effects   (d) a and c

2. To be *Code*-compliant, listed or labeled equipment must be installed and used in accordance with any instructions included in the _____.

   (a) catalog   (b) product   (c) listing or labeling   (d) all of these

3. In the *NEC*, conductors must be _____ unless otherwise provided.

   (a) bare   (b) stranded   (c) copper   (d) aluminum

4. Conductor sizes are expressed in American Wire Gage (AWG) or in _____.

   (a) in.   (b) circular mils   (c) sq in.   (d) AWG

5. All wiring must be installed so that the completed system will be free from _____, other than required or permitted in Article 250.

   (a) short circuits   (b) grounds   (c) a and b   (d) none of these

6. A wiring method included in the *Code* is recognized as being a(n) _____ wiring method.

   (a) expensive   (b) efficient   (c) suitable   (d) cost-effective

7. Equipment intended to break current at other than fault levels must have an interrupting rating at nominal circuit voltage sufficient for the current that must be interrupted.

   (a) True   (b) False

(• Indicates that 75% or fewer exam takers get the question correct)

8. Circuit-protective devices are used to clear a fault without the occurrence of extensive damage to the electrical components of the circuit. Faults can occur between two or more of the _____ or between any circuit conductor and the grounding conductor or enclosing metal raceway.

   (a) bonding jumpers   (b) grounding jumpers   (c) wiring harnesses   (d) circuit conductors

9. The _____ of a circuit must be so selected and coordinated as to permit the circuit protective devices to clear a fault without extensive damage to the electrical components of the circuit.

   (a) overcurrent protective devices   (b) total circuit impedance
   (c) component short-circuit current ratings   (d) all of these

10. Unless identified for use in the operating environment, no conductors or equipment can be _____ having a deteriorating effect on the conductors or equipment.

    (a) located in damp or wet locations   (b) exposed to fumes, vapors, or gases
    (c) exposed to liquids or excessive temperatures   (d) all of these

11. •Equipment approved for use in dry locations only must be protected against permanent damage from the weather during _____.

    (a) design   (b) building construction   (c) inspection   (d) none of these

12. Some cleaning and lubricating compounds contain chemicals that cause severe deteriorating reactions with plastics.

    (a) True   (b) False

13. The *NEC* requires that electrical work be installed _____.

    (a) in a neat and workmanlike manner   (b) under the supervision of a qualified person
    (c) completed before being inspected   (d) all of these

14. Accepted industry workmanship practices are described in ANSI/NECA 1-2000, *Standard Practices for Good Workmanship in Electrical Contracting,* and other ANSI approved installation standards.

    (a) True   (b) False

15. Unused cable or raceway openings in electrical equipment must be _____

    (a) filled with cable clamps or connectors only
    (b) taped over with electrical tape
    (c) repaired only by welding or brazing in a metal slug
    (d) effectively closed by fittings that provide protection substantially equivalent to the wall of the equipment

16. The *Code* prohibits damage to the internal parts of electrical equipment by foreign material such as paint, plaster, cleaners, etc. Precautions must be taken to provide protection from the detrimental effects of paint, plaster, cleaners, etc. on internal parts such as _____.

    (a) busbars   (b) wiring terminals   (c) insulators   (d) all of these

17. For mounting electrical equipment on a masonry wall, it is acceptable to drill a hole in the masonry and drive a wooden plug into the hole, then use sheet rock screws drilled into the wooden plug

    (a) True   (b) False

(• Indicates that 75% or fewer exam takers get the question correct)

18. Electrical equipment that depends on _____ for cooling of exposed surfaces must be installed so that airflow over such surfaces is not prevented by walls or by adjacent installed equipment.

    (a) outdoor air
    (b) natural circulation of air and convection
    (c) artificial cooling and circulation
    (d) magnetic induction

19. Many terminations and equipment are marked with _____.

    (a) an etching tool
    (b) a removable label
    (c) a tightening torque
    (d) the manufacturer's initials

20. Connection of conductors to terminal parts must ensure a thoroughly good connection without damaging the conductors and must be made by means of _____.

    (a) solder lugs
    (b) pressure connectors
    (c) splices to flexible leads
    (d) any of these

21. Soldered splices must first be spliced or joined so as to be mechanically and electrically secure without solder and then be soldered.

    (a) True
    (b) False

22. The temperature rating associated with the ampacity of a _____ must be so selected and coordinated so as not to exceed the lowest temperature rating of any connected termination, conductor, or device.

    (a) terminal
    (b) conductor
    (c) device
    (d) all of these

23. Conductors must have their ampacity determined using the _____ column of Table 310.16 for circuits rated 100A or less or marked for 14 AWG through 1 AWG conductors, unless the equipment terminals are listed for use with higher temperature rated conductors.

    (a) 60°C
    (b) 75°C
    (c) 30°C
    (d) 90°C

24. For circuits rated 100A or less, when the equipment terminals are listed for use with 75°C conductors, the _____ column of Table 310.16 must be used to determine the ampacity of THHN conductors installed.

    (a) 60°C
    (b) 75°C
    (c) 30°C
    (d) 90°C

25. •What size THHN conductor is required for a 50A circuit if the equipment is listed and identified for use with a 75°C conductor? Tip: Table 310.16 lists conductor ampacities.

    (a) 10 AWG
    (b) 8 AWG
    (c) 6 AWG
    (d) all of these

26. Conductors must have their ampacity determined using the _____ column of Table 310.16 for circuits rated over 100A or marked for conductors larger than 1 AWG, unless the equipment terminals are listed for use with higher temperature rated conductors.

    (a) 60°C
    (b) 75°C
    (c) 30°C
    (d) 90°C

27. Separately-installed pressure connectors must be used with conductors at the _____ not exceeding the ampacity at the listed and identified temperature rating of the connector.

    (a) voltages
    (b) temperatures
    (c) listings
    (d) ampacities

(• Indicates that 75% or fewer exam takers get the question correct)

28. On a _____ secondary where the midpoint of one phase winding is grounded, the phase conductor having the higher voltage-to-ground must be identified by an outer finish that is orange in color, or by tagging or other effective means. Such identification must be placed at each point where a connection is made if the grounded neutral conductor is also present.

    (a) 1Ø, 3-wire
    (b) 3Ø, 4-wire delta-connected
    (c) 3Ø, 4-wire wye-connected
    (d) 3Ø, 3-wire delta-connected

29. The high leg (wild leg) of a 3Ø, 4-wire delta-connected system must be identified by using _____.

    (a) an outer finish that is red in color or by other effective means
    (b) an outer finish that is orange in color or by other effective means
    (c) permanent lettering on the conductor installed by the manufacturer of the wire
    (d) this is no longer required

30. Identification of the high leg of a 3Ø, 4-wire delta connected system is required _____.

    (a) at the service disconnect only
    (b) at each point on the system where a connection is made if the equipment grounding conductor is also present
    (c) at each point on the system where a connection is made if the grounding electrode conductor is also present
    (d) at each point on the system where a connection is made if the grounded neutral conductor is also present

31. Switchboards, panelboards, industrial control panels, meter socket enclosures, and motor control centers that are in other than dwelling occupancies and are likely to require examination, adjustment, servicing, or maintenance while _____ must be field marked to warn qualified persons of potential electric arc flash hazards.

    (a) being installed
    (b) energized
    (c) de-energized
    (d) in fault condition

32. Switchboards, panelboards, industrial control panels, meter socket enclosures, and motor control centers in commercial and industrial occupancies that are likely to require _____ while energized must be field marked to warn qualified persons of the danger associated with an arc flash from line-to-line or ground faults.

    (a) examination
    (b) adjustment
    (c) servicing or maintenance
    (d) a, b, or c

33. The manufacturer's name, trademark, or other descriptive marking must be placed on all electric equipment. Where required by the *Code*, markings such as voltage, current, wattage, or other ratings must be provided with sufficient durability to withstand _____.

    (a) the voltages encountered
    (b) painting and other finishes applied
    (c) the environment involved
    (d) lack of planning by the installer

34. Each disconnecting means must be legibly marked to indicate its purpose unless located and arranged so _____.

    (a) that they can be locked out and tagged
    (b) they are not readily accessible
    (c) the purpose is evident
    (d) that they operate at less than 300 volts-to-ground

35. Sufficient access and _____ must be provided and maintained about all electrical equipment to permit ready and safe operation and maintenance of such equipment.

    (a) ventilation
    (b) cleanliness
    (c) circulation
    (d) working space

36. Enclosures housing electrical apparatus that are controlled by a lock are considered _____ to qualified persons.

    (a) readily accessible
    (b) accessible
    (c) available
    (d) none of these

(• Indicates that 75% or fewer exam takers get the question correct)

37. Working-space distances for enclosed live parts must be measured from the _____ of equipment or apparatus, if such are enclosed.

    (a) enclosure   (b) opening   (c) a or b   (d) none of these

38. A minimum of _____ of working clearance is required to live parts operating at 300 volts-to-ground, where there are exposed live parts on one side and no live or grounded parts on the other side.

    (a) 2 ft   (b) 3 ft   (c) 4 ft   (d) 6 ft

39. The minimum working clearance on a circuit that is 120V to ground, with exposed live parts on one side and no live or grounded parts on the other side of the working space, is _____.

    (a) 1 ft   (b) 3 ft   (c) 4 ft   (d) 6 ft

40. •The required working clearance for access to live parts operating at 300V to ground, where there are exposed live parts on one side and grounded parts on the other side, is _____ according to Table 110.26(A).

    (a) 3 ft   (b) 3½ ft   (c) 4 ft   (d) 4½ ft

41. Concrete, brick, or tile walls are considered as _____, as it applies to working-space requirements.

    (a) inconsequential   (b) in the way   (c) grounded   (d) none of these

42. The dimension of working clearance for access to live parts operating at 300V, nominal-to-ground, where there are exposed live parts on both sides of the workspace is _____ according to Table 110.26(A)(1).

    (a) 3 ft   (b) 3½ ft   (c) 4 ft   (d) 4½ ft

43. The working space in front of the electric equipment must not be less than _____ wide, or the width of the equipment, whichever is greater.

    (a) 15 in.   (b) 30 in.   (c) 40 in.   (d) 60 in.

44. Equipment such as raceways, cables, wireways, cabinets, panels, etc. can be located above or below other electrical equipment when the associated equipment does not extend more than _____ from the front of the electrical equipment.

    (a) 3 in.   (b) 6 in.   (c) 12 in.   (d) 30 in.

45. When normally-enclosed live parts are exposed for inspection or servicing, the working space, if in a passageway or general open space, must be suitably _____.

    (a) accessible   (b) guarded   (c) open   (d) enclosed

46. Working space cannot be used for _____.

    (a) storage   (b) raceways   (c) lighting   (d) accessibility

47. For equipment rated 1,200A or more that contains overcurrent devices, switching devices, or control devices, at least one entrance, measuring not less than 24 in. wide and 6½ ft high, must be provided at each end of the working space. Where the entrance to the working space has a personnel door, the door _____.

    (a) must open either in or out with simple pressure and must not have any lock
    (b) must open in the direction of egress and be equipped with panic hardware or other devices so the door can open under simple pressure
    (c) must be removed
    (d) must be equipped with an electronic opener

(• Indicates that 75% or fewer exam takers get the question correct)

48. For equipment rated 1,200A or more that contains overcurrent devices, switching devices, or control devices, there must be one entrance to the required working space not less than 24 in. wide and 6 ft 6 in. high at each end of the working space. Where the depth of the working space is twice that required by 110.26(A)(1), _____ entrance(s) are permitted.

    (a) one   (b) two   (c) three   (d) none of these

49. Illumination must be provided for all working spaces about service equipment, switchboards, panelboards, and motor control centers _____.

    (a) over 600 volts
    (b) located indoors
    (c) Rated 1,200 amperes or more
    (d) Using automatic means of control

50. The minimum headroom of working spaces about motor control centers must be _____.

    (a) 3 ft   (b) 5 ft   (c) 6 ft   (d) 6½ ft

51. The minimum headroom for working spaces about service equipment, switchboards, panelboards, or motor control centers must be 6½ ft, except for service equipment or panelboards in existing dwelling units that do not exceed 200A.

    (a) True   (b) False

52. The dedicated equipment space for electrical equipment that is required for panelboards is measured from the floor to a height of _____ above the equipment, or to the structural ceiling, whichever is lower.

    (a) 3 ft   (b) 6 ft   (c) 12 ft   (d) 30 ft

53. •Heating, cooling, or ventilating equipment (including ducts) that service the electrical room or space cannot be installed in the dedicated space above a panelboard or switchboard.

    (a) True   (b) False

54. The dedicated space above a panelboard extends to a dropped or suspended ceiling, which is considered a structural ceiling.

    (a) True   (b) False

(• Indicates that 75% or fewer exam takers get the question correct)

Questions for Chapter 2— Wiring and Protection

Article 200 Use and Identification of Grounded Neutral Conductors

Article Overview

This article contains the requirements for identification of the grounded neutral conductor and its terminals. If you go back to Article 100, you will see that the *grounded conductor* isn't the same as the *grounding conductor*. Make sure you clearly understand the difference between the two before you begin your study of Article 200.

This article isn't very long, and it's not very complicated, and following these requirements can mean the difference between a safe installation and an electrocution hazard. The illustrations in the *Understanding the NEC, Volume 1* textbook will help you form mental pictures of the key points.

Grounded Conductor. Most electrical power supplies have one output terminal of the power supply bonded to the case of the power supply (system bonding jumper). The conductor that is connected to this grounded terminal is called a "grounded conductor."

Neutral Conductor. The IEEE dictionary defines a neutral conductor as the conductor with an equal potential difference between it and the other output conductors of a 3- or 4-wire system. Therefore, a neutral conductor is the white/gray wire of a 3-wire single-phase 120/240V system, or of a 4-wire three-phase 120/208V or 277/480V system.

Since a neutral conductor must have equal potential between it and all ungrounded conductors in a 3- or 4-wire system, the white wire of a 2-wire circuit, and the white wire from a 4-wire three-phase 120/240V delta-connected system are not neutral conductors—they're grounded conductors.

> **Author's Comment:** The electrical trade industry typically uses the term "neutral," when referring to the white/gray wire. However, the proper term for this conductor is "grounded conductor."

Technically, it's improper to call a "grounded conductor" a "neutral conductor" or "neutral wire" when it's not truly a neutral conductor, but this is a long-standing industry practice. For the purpose of this workbook this conductor will be called a grounded neutral conductor. That should keep *most people happy.*

Questions

1. Article 200 contains the requirements for _____.

    (a) identification of terminals
    (b) grounded neutral conductors in premises wiring systems
    (c) identification of grounded neutral conductors
    (d) all of these

(• Indicates that 75% or fewer exam takers get the question correct)

2. Premises wiring must not be electrically connected to a supply system unless the supply system contains, for any grounded neutral conductor of the interior system, a corresponding conductor that is ungrounded.

   (a) True   (b) False

3. An insulated grounded neutral conductor of _____ or smaller must be identified by a continuous white or gray outer finish, or by three continuous white stripes on other than green insulation along its entire length.

   (a) 3 AWG   (b) 4 AWG   (c) 6 AWG   (d) 8 AWG

4. Grounded neutral conductors _____ and larger must be identified by a continuous white or gray outer finish along their entire length, by three continuous white stripes along their entire length, or by distinctive white or gray markings such as tape, paint, or other effective means at their terminations.

   (a) 10 AWG   (b) 8 AWG   (c) 6 AWG   (d) 4 AWG

5. Grounded neutral conductors larger than 6 AWG must be identified by _____.

   (a) a continuous white or gray outer finish along their entire length
   (b) three continuous white stripes along their entire length
   (c) distinctive white or gray tape or paint at terminations
   (d) a, b, or c

6. Where grounded neutral conductors of different wiring systems are installed in the same raceway, cable, or enclosure, each grounded neutral conductor must be identified by a different one of the acceptable methods in order to distinguish the grounded neutral conductors of each system from the other.

   (a) True   (b) False

7. Where grounded neutral conductors of different wiring systems are installed in the same raceway, cable, or enclosure, each grounded neutral conductor must be identified in a manner that makes it possible to distinguish the grounded neutral conductors for each system. This means of identification must be_____.

   (a) permanently posted at each branch-circuit panelboard
   (b) posted inside each junction box where both system neutrals are present
   (c) done using a listed labeling technique
   (d) all of these

8. The white conductor within a cable can be used for the ungrounded (hot) conductor, but the white conductor must be permanently reidentified to indicate its use as an ungrounded (hot) conductor at each location where the conductor is visible and accessible. Identification must _____.

   (a) be by painting or other effective means
   (b) be a color other than white, gray, or green
   (c) both a and b
   (d) none of these

9. A cable containing an insulated conductor with a white outer finish can be used for single pole, 3-way or 4-way switch loops, if it is permanently reidentified by painting or other effective means at its termination, and at each location where the conductor is visible and accessible.

   (a) True   (b) False

10. The identification of _____ to which a grounded neutral conductor is to be connected must be substantially white in color.

    (a) wire connectors   (b) circuit breakers   (c) terminals   (d) ground rods

(• Indicates that 75% or fewer exam takers get the question correct)

11. Receptacles, polarized attachment plugs, and cord connectors for plugs and polarized plugs must have the terminal intended for connection to the grounded neutral conductor identified. Identification must be by a metal or metal coating that is substantially _____ in color, or by the word white or the letter W located adjacent to the identified terminal.

    (a) green (b) white (c) gray (d) b or c

12. The screw shell of a luminaire or lampholder must be connected to the _____

    (a) grounded neutral conductor
    (b) ungrounded conductor
    (c) equipment grounding conductor
    (d) forming shell terminal

13. No _____ can be attached to any terminal or lead so as to reverse designated polarity.

    (a) grounded neutral conductor (b) grounding conductor (c) ungrounded conductor (d) grounding connector

(• Indicates that 75% or fewer exam takers get the question correct)

Article 210 Branch Circuits

Article Overview

This article contains the requirements for branch circuits, such as conductor sizing and identification, GFCI receptacle protection, and receptacle and lighting outlet requirements. It consists of three parts:

- PART I. GENERAL PROVISIONS
- PART II. BRANCH-CIRCUIT RATINGS
- PART III. REQUIRED OUTLETS

Table 210.2 of this article identifies specific purpose branch circuits. When people complain that the Code "buries stuff in the last few chapters and doesn't provide you with any way of knowing where to find things," that is because they didn't pay attention to this table.

The following Sections and Tables contain a few key items to spend extra time on as you study Article 210:

- 210.4. Multiwire Branch Circuits. The conductors of these circuits must originate from the same panel. These circuits can supply only line-to-neutral loads.

- 210.8. GFCI Protected Receptacles. Crawl spaces, unfinished basements, and boathouses are just some of the eight locations that require GFCI protection.

- 210.11. Branch Circuits Required. With three subheadings, 210.11 gives summarized requirements for the number of branch circuits in a given system, states that a load computed on a VA/area basis must be evenly proportioned, and covers rules for dwelling units.

- 210.12. Arc-Fault Circuit-Interrupter Protection. An AFCI isn't a GFCI, though combination units do exist. The purpose of an AFCI (trips at 30 mA) is to protect equipment. The purpose of a GFCI (trips at 4 to 6 mA) is to protect people.

- 210.19. Conductors—Minimum Ampacity and Size. This gets complicated in a hurry, but the Understanding the NEC, Volume 1 textbook provides additional guidance to help you through it.

- Table 210.21(B)(2) shows that the maximum load on a given circuit is 80 percent of the receptacle rating and circuit rating. More about the implications of this are explained in the textbook.

- 210.23. Permissible Loads. This is intended to prevent a circuit overload from occurring just because someone plugs in a lamp or vacuum cleaner. The Understanding the NEC, Volume 1 textbook will show you how to conform.

- 210.52. Dwelling Unit Receptacle Outlets. An area rife with confusion is receptacle spacing. The textbook cuts through the confusion, and you'll understand the meaning of 210.52 and how to apply it correctly.

The rest of the material is also important. But mastering these key items will give you a decided edge in your ability to do work free of Code violations.

Questions

1. The rating of a branch circuit is determined by the rating of the _____.

    (a) ampacity of the largest device connected to the circuit
    (b) average of the ampacity of all devices
    (c) branch-circuit overcurrent protection
    (d) ampacity of the branch circuit conductors according to Table 310.16

(• Indicates that 75% or fewer exam takers get the question correct)

2. •Multiwire branch circuits must _____.

   (a) supply only line-to-neutral loads
   (b) not be allowed in dwelling units
   (c) have their conductors originate from different panelboards
   (d) none of these

3. When more than one nominal voltage system exists in a building, each ungrounded system conductor must be identified by system. The means of identification must be permanently posted at each branch-circuit panelboard.

   (a) True   (b) False

4. Where more than one nominal voltage system exists in a building, each ungrounded conductor of a branch circuit, where accessible, must be identified by system. The identification can be _____ and must be permanently posted at each branch-circuit panelboard.

   (a) color-coding   (b) phase tape   (c) tagging   (d) any of these

5. Where more than one nominal voltage system exists in a building, each _____ conductor of a branch circuit, where accessible, must be identified by system.

   (a) grounded   (b) ungrounded   (c) grounding   (d) all of these

6. In dwelling units, the voltage between conductors that supply the terminals of _____ must not exceed 120V, nominal.

   (a) luminaires
   (b) cord-and-plug connected loads of 1,440 VA, nominal, or less
   (c) cord-and-plug connected loads of more than ¼ hp
   (d) a and b

7. A branch-circuit voltage that exceeds 277 volts-to-ground and does not exceed 600V between conductors is used to wire the auxiliary equipment of electrical discharge lamps mounted on poles. The minimum height of these luminaires must not be less than _____.

   (a) 31 ft   (b) 15 ft   (c) 18 ft   (d) 22 ft

8. All ungrounded (hot) conductors from two or more branch circuits terminating on multiple devices or equipment on the same yoke must have a means to be disconnected simultaneously in _____ occupancies.

   (a) dwelling unit   (b) commercial   (c) industrial   (d) all of these

9. Where two or more branch circuits supply devices or equipment on the same yoke, a means to disconnect simultaneously all ungrounded (hot) conductors that supply those devices or equipment must be provided _____.

   (a) at the point where the branch circuit originates
   (b) at the location of the device or equipment
   (c) at the point where the feeder originates
   (d) none of these

10. All 15 and 20A, 125V single-phase receptacles installed in bathrooms of _____ must have ground-fault circuit-interrupter (GFCI) protection for personnel.

    (a) guest rooms in hotels/motels
    (b) dwelling units
    (c) office buildings
    (d) all of these

11. GFCI protection for personnel is required for all 15 and 20A, 125V single-phase receptacles installed in a dwelling unit _____.

    (a) attic   (b) garage   (c) laundry room   (d) all of these

(• Indicates that 75% or fewer exam takers get the question correct)

12. GFCI protection is required for all 15 and 20A, 125V single-phase receptacles in accessory buildings that have a floor located at or below grade level not intended as _____ and limited to storage areas, work areas, or similar use.

    (a) habitable   (b) finished   (c) a or b   (d) none of these

13. •A _____ receptacle without GFCI protection can be located in a dwelling unit garage to supply one appliance, which is not easily moved, if the receptacle is located within the dedicated space for the appliance.

    (a) multioutlet   (b) duplex   (c) single   (d) none of these

14. GFCI protection for personnel is required for fixed electric snow melting or deicing equipment receptacles that are not readily accessible and are supplied by a dedicated branch circuit.

    (a) True   (b) False

15. All 15 and 20A, 125V single-phase receptacles installed in crawl spaces at or below grade level and in _____ of dwelling units must have GFCI protection for personnel.

    (a) unfinished attics   (b) finished attics   (c) unfinished basements   (d) finished basements

16. GFCI protection for personnel is required for all 15 and 20A, 125V single-phase receptacles installed to serve the countertop surfaces in dwelling unit kitchens.

    (a) True   (b) False

17. GFCI protection is required for all 15 and 20A, 125V single-phase receptacles located within an arc measurement of 6 ft from the dwelling unit _____.

    (a) laundry sink   (b) utility sink   (c) wet bar sink   (d) all of these

18. All 15 and 20A, 125V single-phase receptacles installed in dwelling unit boathouses must have GFCI protection for personnel.

    (a) True   (b) False

19. GFCI protection for personnel is required for all 15 and 20A, 125V single-phase receptacles installed _____ of commercial, industrial, and all other nondwelling occupancies.

    (a) in storage rooms   (b) in equipment rooms   (c) in warehouses   (d) in bathrooms

20. GFCI protection for personnel is required for all 15 and 20A, 125V single-phase receptacles installed on rooftops in other than dwelling units, including those for fixed electric snow melting or deicing equipment.

    (a) True   (b) False

21. All 15 and 20A, 125V single-phase receptacles _____ of commercial occupancies must have GFCI protection for personnel.

    (a) in bathrooms   (b) on rooftops   (c) in kitchens   (d) all of these

22. In locations other than dwelling units, a kitchen _____.

    (a) is required to have GFCI protection on all 15 and 20A, 125V single-phase receptacles
    (b) includes a sink
    (c) includes permanent facilities for food preparation and cooking
    (d) all of these

(• Indicates that 75% or fewer exam takers get the question correct)

23. In other than dwelling units, GFCI protection is required _____.

    (a) for outdoor 15 and 20A, 125V single-phase receptacles accessible to the public
    (b) at an accessible location for HVAC equipment
    (c) both a and b
    (d) neither a nor b

24. Ground-fault circuit-interrupter protection for personnel must be provided for outlets that supply boat hoists installed in dwelling unit locations and supplied by a 15 or 20A, 120V branch circuit.

    (a) True   (b) False

25. Where the load is computed on volt-amperes per square meter or square foot basis, the wiring system up to and including the branch-circuit _____ must be provided to serve not less than the calculated load.

    (a) wiring   (b) protection   (c) panelboard(s)   (d) all of these

26. Two or more _____, 120V small-appliance branch circuits must be provided to supply power for the receptacle outlets in the dwelling unit kitchen, dining room, breakfast room, pantry, or similar dining areas.

    (a) 15A   (b) 20A   (c) 30A   (d) either 20A or 30A

27. There must be a minimum of one _____ branch circuit for the laundry outlet(s) in a dwelling unit.

    (a) 15A   (b) 20A   (c) 30A   (d) b and c

28. An individual 20A circuit is permitted to supply power to a single dwelling unit bathroom for receptacle outlet(s) and other equipment within the same bathroom.

    (a) True   (b) False

29. All branch circuits that supply 15 and 20A, 125V single-phase outlets installed in dwelling unit bedrooms must be protected by a(n) _____ listed to provide protection of the entire branch circuit.

    (a) AFCI   (b) GFCI   (c) a and b   (d) none of these

30. All 15 or 20A, 120V branch circuits that supply outlets in dwelling unit bedrooms must be AFCI protected by a listed arc-fault circuit interrupter of the combination type after January 1, 2008

    (a) True   (b) False

31. The location of the arc-fault circuit interrupter can be at other than the origination of the branch circuit if _____.

    (a) the arc-fault circuit interrupter is installed within 6 ft of the branch-circuit overcurrent device
    (b) the circuit conductors up to the arc-fault circuit interrupter are in a metal raceway or a cable with a metallic sheath
    (c) both a and b
    (d) none of these

32. _____ provided with permanent provisions for cooking must have branch circuits and outlets installed to meet the rules for dwelling units.

    (a) Guest rooms   (b) Guest suites   (c) Commercial kitchens   (d) a and b

(• Indicates that 75% or fewer exam takers get the question correct)

33. Branch-circuit conductors that supply a continuous load, or any combination of continuous and non-continuous loads, must have an ampacity of not less than 125 percent of the continuous load, plus 100 percent of the noncontinuous load.

    (a) True   (b) False

34. The recommended maximum total voltage drop on both the feeder and branch-circuit conductors combined is _____ percent.

    (a) 3   (b) 2   (c) 5   (d) 4.6

35. The grounded neutral conductor of a 3-wire branch circuit supplying a household electric range is permitted to be smaller than the ungrounded conductors when the maximum demand of a range of 8.75 kW or more rating has been computed according to Column C of Table 220.19. However, the ampacity of the grounded neutral conductor must not be less than _____ percent of the branch-circuit rating and not be smaller than _____ AWG.

    (a) 50, 6   (b) 70, 6   (c) 50, 10   (d) 70, 10

36. Where a branch circuit supplies continuous loads, or any combination of continuous and noncontinuous loads, the rating of the overcurrent device must not be less than the noncontinuous load plus 125 percent of the continuous load.

    (a) True   (b) False

37. •A single receptacle installed on an individual branch circuit must be rated at least _____ percent of the rating of the branch circuit.

    (a) 50   (b) 60   (c) 90   (d) 100

38. When connected to a branch circuit supplying _____ or more receptacles or outlets, a receptacle must not supply a total cord-and-plug connected load in excess of the maximum specified in Table 210.21(B)(2).

    (a) two   (b) three   (c) four   (d) five

39. What is the maximum cord-and-plug connected load permitted on a 15A receptacle that is supplied by a 20A circuit supplying multiple outlets?

    (a) 12A   (b) 16A   (c) 20A   (d) 24A

40. •If a 20A branch circuit supplies multiple 125V receptacles, the receptacles must have an ampere rating of no less than _____.

    (a) 10A   (b) 15A   (c) 20A   (d) 30A

41. It is permitted to base the _____ rating of a range receptacle on a single range demand load specified in Table 220.19.

    (a) circuit   (b) voltage   (c) ampere   (d) resistance

42. The total rating of utilization equipment fastened in place, other than luminaires, must not exceed _____ percent of the branch-circuit ampere rating where the circuit also supplies receptacles for cord-and-plug connected equipment not fastened in place and/or lighting units.

    (a) 50   (b) 75   (c) 100   (d) 125

43. Multioutlet circuits rated 15 or 20A can supply fixed appliances (utilization equipment fastened in place) as long as the fixed appliances do not exceed _____ percent of the circuit rating.

    (a) 125   (b) 100   (c) 75   (d) 50

(• Indicates that 75% or fewer exam takers get the question correct)

44. _____ in dwelling units must supply only loads within that dwelling unit or loads associated only with that dwelling unit.

    (a) Service-entrance conductors
    (b) Ground-fault protection
    (c) Branch circuits
    (d) none of these

45. A cord connector on a permanently installed cord pendant is considered a receptacle outlet.

    (a) True
    (b) False

46. Receptacle outlets installed for a specific appliance in a dwelling unit, such as a clothes washer, dryer, range, or refrigerator, must be within _____ of the intended location of the appliance.

    (a) sight
    (b) 6 ft
    (c) 3 ft
    (d) readily accessible, no maximum distance

47. In a dwelling unit, the minimum required receptacle outlets must be in addition to receptacle outlets that are _____.

    (a) part of a luminaire or appliance
    (b) located within cabinets or cupboards
    (c) located more than 5½ ft above the floor
    (d) all of these

48. When applying the general provisions for receptacle spacing to the rooms of a dwelling unit, which require receptacles in the wall space, no point along the floor line in any wall space of a dwelling unit may be more than _____ from an outlet.

    (a) 12 ft
    (b) 10 ft
    (c) 8 ft
    (d) 6 ft

49. In dwelling units, when determining the spacing of general-use receptacles, _____ on exterior walls are not considered wall space.

    (a) fixed panels
    (b) fixed glass
    (c) sliding panels
    (d) all of these

50. In a dwelling unit, each wall space of _____ or wider requires a receptacle.

    (a) 2 ft
    (b) 3 ft
    (c) 4 ft
    (d) 5 ft

51. Receptacle outlets in floors are not counted as part of the required number of receptacle outlets to service dwelling unit wall spaces unless they are located _____ the wall.

    (a) within 6 in. of
    (b) within 12 in. of
    (c) within 18 in. of
    (d) close to

52. In dwelling units, outdoor receptacles can be connected to one of the 20A small-appliance branch circuits.

    (a) True
    (b) False

53. The small-appliance branch circuits can supply the _____ as well as the kitchen.

    (a) dining room
    (b) refrigerator
    (c) breakfast room
    (d) all of these

54. A receptacle connected to one of the small-appliance branch circuits can be used to supply an electric clock.

    (a) True
    (b) False

55. A receptacle connected to a small-appliance circuit can supply gas-fired ranges, ovens, or counter-mounted cooking units.

    (a) True
    (b) False

(• Indicates that 75% or fewer exam takers get the question correct)

56. Receptacles installed in a kitchen to serve countertop surfaces must be supplied by not fewer than _____ small-appliance branch circuits.

    (a) one   (b) two   (c) three   (d) no minimum

57. Two 20A small-appliance branch circuits can supply more than one kitchen in a dwelling.

    (a) True   (b) False

58. A receptacle outlet must be installed at each wall counter space that is 12 in. or wider so that no point along the wall line is more than _____, measured horizontally, from a receptacle outlet in that space.

    (a) 10 in.   (b) 12 in.   (c) 16 in.   (d) 24 in.

59. A receptacle outlet must be installed in dwelling units for every kitchen and dining area countertop space _____, and no point along the wall line can be more than 2 ft, measured horizontally, from a receptacle outlet in that space.

    (a) wider than 10 in.   (b) wider than 3 ft   (c) 18 in. or wider   (d) 12 in. or wider

60. At least one receptacle outlet must be installed at each peninsular countertop or island not containing a sink or range top, having a long dimension of _____ in. or greater, and a short dimension of _____ in. or greater.

    (a) 12, 24   (b) 24, 12   (c) 24, 48   (d) 48, 24

61. One receptacle outlet must be installed at each island or peninsular countertop space with a long dimension of 2 ft or greater, and a short dimension of 12 in. or greater. When breaks occur in countertop spaces for appliances, sinks, etc., there is never a need for more than one receptacle outlet.

    (a) True   (b) False

62. For the purpose of determining the placement of receptacles in a dwelling unit kitchen, a(n) _____ countertop is measured from the connecting edge.

    (a) island   (b) usable   (c) peninsular   (d) cooking

63. When breaks occur in dwelling unit kitchen countertop spaces for ranges, refrigerators, sinks, etc., each countertop surface is considered a separate counter space for determining receptacle placement.

    (a) True   (b) False

64. Kitchen and dining room countertop receptacle outlets in dwelling units must be installed above the countertop surface, and not more than ___ above the countertop.

    (a) 12 in.   (b) 20 in.   (c) 24 in.   (d) none of these

65. Receptacle outlets can be installed below the countertop surface in dwelling units when necessary for the physically impaired, or if there is no means available to mount a receptacle above an island or peninsular countertop.

    (a) True   (b) False

66. The required receptacle for a dwelling unit countertop surface can be mounted a maximum height of _____ above a dwelling unit kitchen counter surface.

    (a) 10 in.   (b) 12 in.   (c) 18 in.   (d) 20 in.

(• Indicates that 75% or fewer exam takers get the question correct)

67. In dwelling units, at least one wall receptacle outlet must be installed in bathrooms within _____ of the outside edge of each basin. The receptacle outlet must be located on a wall or partition that is adjacent to the basin or basin countertop.

    (a) 12 in. (b) 18 in. (c) 24 in. (d) 36 in.

68. In dwelling units, the required wall receptacle outlet is allowed to be installed on the side or front of the basin cabinet if no lower than _____ below the countertop.

    (a) 12 in. (b) 18 in. (c) 24 in. (d) 36 in.

69. A one-family or two-family dwelling unit requires a minimum of _____ GFCI receptacle(s) to be installed outdoors.

    (a) zero (b) one (c) two (d) three

70. At least one receptacle outlet accessible from grade level and not more than _____ above grade must be installed at each dwelling unit of a multifamily dwelling located at grade level and provided with individual exterior entrance/egress.

    (a) 3 ft (b) 6½ ft (c) 8 ft (d) 24 in.

71. A receptacle outlet for the laundry is not required in a dwelling unit in a multifamily building when laundry facilities available to all building occupants are provided on the premises.

    (a) True (b) False

72. For a one-family dwelling, at least one receptacle outlet is required in each _____.

    (a) basement
    (b) attached garage
    (c) detached garage with electric power
    (d) all of these

73. Where a portion of the dwelling unit basement is finished into one or more habitable rooms, each separate unfinished portion must have a receptacle outlet installed.

    (a) True (b) False

74. Hallways in dwelling units that are _____ long or longer require a receptacle outlet.

    (a) 12 ft (b) 10 ft (c) 8 ft (d) 15 ft

75. Guest rooms in hotels, motels, and similar occupancies without permanent provisions for cooking must have receptacle outlets installed in accordance with 210.52(A) and 210.52(D).

    (a) True (b) False

76. Guest rooms or guest suites provided with permanent provisions for _____ must have receptacle outlets installed in accordance with all of the applicable requirements for a dwelling unit in accordance with 210.52.

    (a) whirlpool tubs (b) bathing (c) cooking (d) internet access

77. Receptacles installed behind a bed in the guest rooms in hotels and motels must be located so as to prevent the bed from contacting an attachment plug, or the receptacle must be provided with a suitable guard.

    (a) True (b) False

(• Indicates that 75% or fewer exam takers get the question correct)

78. The number of receptacle outlets for guest rooms in hotels and motels must not be less than that required for a dwelling unit, in accordance with 210.52(A). These receptacles can be located to be convenient for permanent furniture layout, but lesson fewer than _____ receptacle outlets must be readily accessible

    (a) 4 (b) 2 (c) 6 (d) 1

79. A 15 or 20A, 125V, single-phase receptacle outlet must be installed at an accessible location for the servicing of heating, air-conditioning, and refrigeration equipment. The receptacle must be located on the same level and within _____ of the heating, air-conditioning, and refrigeration equipment.

    (a) 10 ft (b) 15 ft (c) 20 ft (d) 25 ft

80. A 15 or 20A, 125V, single-phase receptacle outlet must be located within 25 ft of heating, air-conditioning, and refrigeration equipment for _____ occupancies.

    (a) dwelling (b) commercial (c) industrial (d) all of these

81. In a dwelling unit, at least _____ wall switch-controlled lighting outlet(s) must be installed in every dwelling unit habitable room and bathroom.

    (a) one (b) three (c) six (d) none of these

82. Which rooms in a dwelling unit must have a switch-controlled lighting outlet?

    (a) Every habitable room (b) Bathrooms (c) Hallways and stairways (d) all of these

83. In _____ rooms other than kitchens and bathrooms of dwelling units, one or more receptacles controlled by a wall switch are permitted in lieu of lighting outlets.

    (a) habitable (b) finished (c) all (d) a and b

84. Lighting outlets can be controlled by occupancy sensors equipped with a _____ that will allow the sensor to function as a wall switch.

    (a) manual override (b) photo cell
    (c) GFCI device (d) selenium controlled rectifier (SCR)

85. In a dwelling unit, illumination from a lighting outlet must be provided at the exterior side of each outdoor entrance or exit that has grade-level access.

    (a) True (b) False

86. When considering lighting outlets in dwelling units, a vehicle door in a garage is considered an outdoor entrance.

    (a) True (b) False

87. Where a lighting outlet(s) is installed for interior stairways, there must be a wall switch at each floor landing that includes an entryway where the stairway between floor levels has four risers or more.

    (a) True (b) False

88. In a dwelling unit, illumination on the exterior side of outdoor entrances or exits that have grade-level access can be controlled by _____.

    (a) home automation devices (b) motion sensors (c) photocells (d) any of these

(• Indicates that 75% or fewer exam takers get the question correct)

89. In a dwelling unit, at least one lighting outlet _____ located at the point of entry to the attic, underfloor space, utility room, and basement must be installed where these spaces are used for storage or contain equipment requiring servicing.

    (a) that is unswitched and   (b) containing a switch   (c) controlled by a wall switch   (d) b or c

90. At least one wall switch-controlled lighting outlet must be installed in every habitable room and bathroom of a guest room or guest suite of hotels, motels, and similar occupancies. A receptacle outlet controlled by a wall switch may be used to meet this requirement in other than _____.

    (a) bathrooms   (b) kitchens   (c) sleeping areas   (d) both a and b

91. For other than dwelling units, a lighting outlet containing a switch or controlled by a wall switch is required near equipment requiring servicing in attics or underfloor spaces, and at least one point of control must be located at the point of entrance to the attic or underfloor space.

    (a) True   (b) False

(• Indicates that 75% or fewer exam takers get the question correct)

Article 215 Feeders

Article Overview

The next logical step up from the branch circuit is the feeder circuit. Consequently, Article 215 follows Article 210. This article covers the rules for installation, minimum size, and ampacity of feeders.

This is a very short article, and that's puzzling at first glance. It might seem feeders would just be "heavier" branch circuits, and that Article 215 should just be another Article 210 but with more stringent requirements. But this isn't the case at all.

If you go back and look at Article 210 again, you'll see it covers many permutations of branch circuits. It also devotes extensive space to dwelling-area branch circuits. Dwelling units don't have many feeders. A multifamily dwelling building will have at least one feeder for each occupancy.

Here's an object lesson in the value of Article 100. Go there now and review the definitions of branch circuit and feeder. Once you've done that, you will understand why Article 215 is so much shorter than Article 210.

Questions

1. The size of the grounded neutral conductor for a feeder must not be smaller than specified in _____.

   (a) Table 250.122   (b) Table 250.66   (c) Table 310.16   (d) Table 430.52

2. The feeder conductor ampacity must not be less than that of the service-entrance conductors where the feeder conductors carry the total load supplied by service-entrance conductors with an ampacity of _____ or less.

   (a) 100A   (b) 60A   (c) 55A   (d) 30A

3. Dwelling unit or mobile home feeder conductors need not be larger than the service conductors and are permitted to be sized according to 310.15(B)(6).

   (a) True   (b) False

4. If required by the authority having jurisdiction, a diagram showing feeder details must be provided _____ of the feeders.

   (a) after the installation   (b) prior to the installation   (c) before the final inspection   (d) diagrams are not required

5. When a feeder supplies _____ in which equipment grounding conductors are required, the feeder must include or provide a grounding means to which the equipment grounding conductors of the branch circuits must be connected.

   (a) equipment disconnecting means   (b) electrical systems
   (c) branch circuits   (d) electric-discharge lighting equipment

6. Ground-fault protection of equipment is required for the feeder disconnect if _____.

   (a) the feeder is rated 1,000A or more
   (b) it is a solidly-grounded wye system
   (c) it is more than 150 volts-to-ground, but not exceeding 600V phase-to-phase
   (d) all of these

(• Indicates that 75% or fewer exam takers get the question correct)

Mike Holt Enterprises, Inc. • www.NECcode.com • 1.888.NEC.Code

7. Ground-fault protection of equipment is not required at the feeder disconnect if ground-fault protection of equipment is provided on the _____ side of the feeder.

(a) load (b) supply (c) service (d) none of these

8. Where the premises wiring system contains feeders supplied from more than one voltage system, each ungrounded (hot) conductor, where accessible, must be identified by the system. Identification can be by _____ or other approved means. Such identification must be permanently posted at each feeder panelboard or similar feeder distribution equipment.

(a) color coding (b) marking tape (c) tagging (d) a, b, or c

(• Indicates that 75% or fewer exam takers get the question correct)

Article 220 Branch-Circuit, Feeder, and Service Calculations

Article Overview

This article provides the requirements for sizing branch circuits, feeders, and services, and for determining the number of receptacles on a circuit and the number of branch circuits required. It consists of five parts:

- PART I. GENERAL
- PART II. BRANCH-CIRCUIT LOAD CALCULATIONS
- PART III. FEEDER AND SERVICE CALCULATIONS
- PART IV. OPTIONAL CALCULATIONS
- PART V. FARM LOAD CALCULATIONS

Part I describes the layout of Article 220 and provides a table of where other types of load calculations can be found in the NEC. Part II provides requirements for branch-circuit calculations and for specific types of branch circuits. Part III provides requirements for feeder and service calculations, just as the title says. Part IV provides some shortcut calculations you can use in place of the more complicated calculations provided in Parts II and III—if your installation meets certain requirements. Part IV covers just what it says, Farm Load Calculations.

The typical electrician is wise to focus on Parts I, II, and III. Whether to do the optional calculations is typically a decision made by the project manager or design engineer. You need to be aware that there can be two right answers when doing the calculations because the NEC allows two different methods.

The cost of improperly applying Article 220 can be staggering. In the best of all possible worlds, the price of misapplication is just an expensive call-back and some rework. In reality, the costs can easily involve catastrophic destruction and the loss of human life.

So study Article 220 carefully. If something doesn't make sense at first, make a note of it and take a short break from your studies. Then go back to that item and read through the explanation again. If you have the Understanding the NEC, Volume 1 textbook, use the illustrations to help you understand. Your learning will really stick if you also consider the why, not just the how.

Questions

1. When computations in Article 220 result in a fraction of an ampere that is less than_____, such fractions can be dropped.

   (a) 0.49 (b) 0.50 (c) 0.51 (d) none of these

2. The 3 VA per-square-foot general lighting load for dwelling units does not include _____.

   (a) open porches
   (b) garages
   (c) unused or unfinished spaces not adaptable for future use
   (d) all of these

3. •When determining the load for luminaires for branch circuits, the load must be based on the _____.

   (a) wattage rating of the luminaire socket
   (b) maximum VA rating of the equipment and lamps
   (c) wattage rating of the lamps
   (d) none of these

4. Where fixed multioutlet assemblies used in other than dwelling units or the guest rooms of hotels or motels are employed, each _____ or fraction thereof of each separate and continuous length of multioutlet assembly must be considered as one outlet of not less than 180 VA capacity where appliances are unlikely to be used simultaneously.

   (a) 5 ft (b) 5½ ft (c) 6 ft (d) 6½ ft

(• Indicates that 75% or fewer exam takers get the question correct)

Mike Holt Enterprises, Inc. • www.NECcode.com • 1.888.NEC.Code

5. For other than dwelling occupancies, banks, or office buildings, each receptacle outlet must be computed at not less than _____ VA for each single or each multiple receptacle on one yoke.

   (a) 1,500   (b) 180   (c) 20   (d) 3

6. A single piece of equipment consisting of a multiple receptacle comprised of _____ or more receptacles must be computed at not less than 90 VA per receptacle.

   (a) 1   (b) 2   (c) 3   (d) 4

7. The 3 VA per square foot general lighting load for dwelling units includes general use receptacles and lighting outlets and no additional load calculations are required for these.

   (a) True   (b) False

8. •The demand factors of Table 220.42 must apply to the computed load of feeders to areas in hospitals, hotels, and motels where the entire lighting is likely to be used at one time, as in operating rooms, ballrooms, or dining rooms.

   (a) True   (b) False

9. For other than dwelling units or guest rooms of hotels or motels, the feeder and service load calculation for track lighting is to be determined at 150 VA for every _____ of track installed.

   (a) 4 ft   (b) 6 ft   (c) 2 ft   (d) none of these

10. •Receptacle loads for nondwelling units, computed in accordance with 220.14(H) and (I), are permitted to be _____.

    (a) added to the lighting loads and made subject to the demand factors of Table 220.42
    (b) made subject to the demand factors of Table 220.44
    (c) made subject to the lighting demand loads of Table 220.12
    (d) a or b

11. The feeder and service conductors for motors must be computed in accordance with Article _____.

    (a) 450   (b) 240   (c) 430   (d) 100

12. The feeder and service load for fixed electric space heating must be computed at _____ percent of the total connected load.

    (a) 125   (b) 100   (c) 80   (d) 200

13. Loads that are computed for dwelling unit small-appliance branch circuits can be included with the _____ load and subject to the demand factors permitted in Table 220.42 for the general lighting load.

    (a) general lighting   (b) feeder   (c) appliance   (d) receptacle

14. When sizing a feeder for the fixed appliance loads in dwelling units, a demand factor of 75 percent of the total nameplate ratings can be applied if there are _____ or more appliances fastened in place on the same feeder (not including washer, dryer, heating, or air conditioning).

    (a) two   (b) three   (c) four   (d) five

15. Using standard load calculations, the feeder demand factor for five household clothes dryers is _____ percent.

    (a) 70   (b) 85   (c) 50   (d) 100

(• Indicates that 75% or fewer exam takers get the question correct)

16. The load for electric clothes dryers in a dwelling unit must be _____ watts or the nameplate rating, whichever is larger, per dryer.

    (a) 1,500   (b) 4,500   (c) 5,000   (d) 8,000

17. To determine the feeder demand load for ten 3 kW household cooking appliances, use _____ of Table 220.19.

    (a) Column A   (b) Column B   (c) Column C   (d) none of these

18. The feeder demand load for four 6 kW cooktops is _____ kW.

    (a) 17   (b) 4   (c) 12   (d) 24

19. The feeder demand load for nine 12 kW ranges is _____.

    (a) 13,000W   (b) 14,700W   (c) 24,000W   (d) 16,000W

20. For identically sized ranges rated more than 12 kW but not more than 27 kW, the maximum demand in column C must be increased by _____ percent of the column C value for each additional kilowatt of rating, or major fraction thereof, by which the rating of individual ranges exceeds 12 kW.

    (a) 125   (b) 10   (c) 5   (d) 80

21. The feeder demand load for nine 16 kW ranges is _____.

    (a) 15,000W   (b) 28,800W   (c) 20,000W   (d) 26,000W

22. The feeder demand load for ranges individually rated more than 8 3/4 kW and of different ratings, but none exceeding 27 kW, is calculated by adding all of the ranges together and dividing by the total number of ranges to find an average value. The column C value for the number of ranges is then increased by _____ percent for each kW or major fraction that the average value exceeds 12 kW.

    (a) 125   (b) 10   (c) 5   (d) 80

23. The demand factors of Table 220.56 apply to space heating, ventilating, or air-conditioning equipment.

    (a) True   (b) False

24. Table 220.56 may be applied to compute the load for thermostatically controlled or intermittently used _____ and other kitchen equipment in a commercial kitchen.

    (a) commercial electric cooking equipment
    (b) dishwasher booster heaters
    (c) water heaters
    (d) all of these

25. When applying the demand factors of Table 220.56, in no case can the feeder or service demand load be less than the sum of _____.

    (a) the total number of receptacles at 180 VA per receptacle outlet
    (b) the VA rating of all of the small appliance circuits combined
    (c) the largest two kitchen equipment loads
    (d) the kitchen heating and air conditioning loads

26. Where it is unlikely that two or more noncoincident loads will be in use simultaneously, it is permissible to use only the _____ loads on at any given time in computing the total load to a feeder.

    (a) smaller of the (b) largest of the (c) difference between the (d) none of these

(• Indicates that 75% or fewer exam takers get the question correct)

27. •The maximum unbalanced feeder load for household electric ranges, wall-mounted ovens, counter-mounted cooking units, and electric dryers must be considered as _____ percent of the load on the ungrounded conductors as determined in accordance with Table 220.55 for ranges and Table 220.54 for dryers.

    (a) 50   (b) 70   (c) 85   (d) 115

28. There must be no reduction in the size of the grounded neutral conductor on _____ type loads.

    (a) dwelling unit   (b) hospital   (c) nonlinear   (d) motel

29. If a dwelling unit is served by a single 1Ø, 3-wire, 120/240V or 120/208V set of service-entrance or feeder conductors with an ampacity of _____ or greater, it is permissible to compute the feeder and service loads in accordance with 220.82 instead of the method specified in Part III of Article 220.

    (a) 100   (b) 125   (c) 150   (d) 175

30. Feeder and service-entrance conductors with demand loads determined by the use of 220.82 are permitted to have the _____ load determined by 220.61.

    (a) feeder   (b) circuit   (c) neutral   (d) none of these

31. Under the optional method for calculating a single-family dwelling, general loads beyond the initial 10 kW are assessed at a _____ percent demand factor.

    (a) 40   (b) 50   (c) 60   (d) 75

32. A demand factor of _____ percent applies to a multifamily dwelling with ten units if the optional calculation method is used.

    (a) 75   (b) 60   (c) 50   (d) 43

33. The calculated load to which the demand factors of Table 220.84 apply must include 3 VA per _____ for general lighting and general-use receptacles.

    (a) inch   (b) foot   (c) square inch   (d) square foot

34. The calculated load to which the demand factors of Table 220.84 apply must include the _____ rating of all appliances that are fastened in place, permanently connected, or located to be on a specific circuit. These include ranges, wall-mounted ovens, counter-mounted cooking units, clothes dryers, water heaters, and space heaters.

    (a) calculated   (b) nameplate   (c) circuit   (d) overcurrent protection

35. In order to use the optional method for calculating a service to a school, the school must be equipped with _____.

    (a) cooking facilities   (b) electric space heating   (c) air-conditioning   (d) b or c

36. Feeder conductors for new restaurants are not required to be of _____ ampacity than the service-entrance conductors.

    (a) greater   (b) lesser   (c) equal   (d) none of these

37. •Service-entrance or feeder conductors whose demand load is determined by the optional calculation, as permitted in 220.88, are not permitted to have the neutral load determined by 220.61.

    (a) True   (b) False

(• Indicates that 75% or fewer exam takers get the question correct)

Article 225 Outside Wiring

Article Overview

This article covers installation requirements for equipment including conductors located outdoors, on or between buildings, poles, and other structures on the premises. It has two parts:

Part I provides a listing of other articles that may provide additional requirements, addresses some general concerns, and briefly covers conductor sizing. Then it addresses conductor support, attachment, and clearances.

Part II addresses how many supplies you can have to a building and how to disconnect them. This includes such things as where to locate the disconnecting means and how to group them.

Questions

1. Open individual conductors must not be smaller than _____ AWG copper for spans up to 50 ft in length and _____ AWG copper for a longer span, unless supported by a messenger wire.

   (a) 10, 8   (b) 6, 8   (c) 6, 6   (d) 8, 8

2. The minimum point of attachment of overhead conductors to a building must in no case be less than _____ above finished grade.

   (a) 8 ft   (b) 10 ft   (c) 12 ft   (d) 15 ft

3. Where a mast is used for overhead conductor support of outside branch circuits and feeders, it must have adequate mechanical strength, or braces or guy wires to support it, to withstand the strain caused by the conductors. Only _____ conductors can be attached to the mast.

   (a) communications   (b) fiber optic   (c) feeder or branch circuit   (d) all of these

4. Overhead conductors must have a minimum of _____ vertical clearance from final grade over residential property and driveways, as well as those commercial areas not subject to truck traffic where the voltage is limited to 300 volts-to-ground.

   (a) 10 ft   (b) 12 ft   (c) 15 ft   (d) 18 ft

5. The minimum clearance for overhead conductors not exceeding 600V that pass over commercial areas subject to truck traffic is _____.

   (a) 10 ft   (b) 12 ft   (c) 15 ft   (d) 18 ft

6. Overhead conductors installed over roofs must have a vertical clearance of _____ above the roof surface.

   (a) 8 ft   (b) 12 ft   (c) 15 ft   (d) 3 ft

7. If a set of 120/240V overhead conductors terminates at a through-the-roof raceway or approved support, with less than 6 ft of these conductors passing over the roof overhang, the minimum clearance above the roof for these conductors is _____.

   (a) 12 in.   (b) 18 in.   (c) 2 ft   (d) 5 ft

(• Indicates that 75% or fewer exam takers get the question correct)

8. The requirement for maintaining a 3 ft vertical clearance from the edge of the roof does not apply to the final conductor span where the conductors are attached to _____.

   (a) a building pole   (b) the side of a building   (c) an antenna   (d) the base of a building

9. Overhead conductors to a building must maintain a vertical clearance of final spans above, or within _____ measured horizontally from the platforms, projections, or surfaces from which they might be reached.

   (a) 3 ft   (b) 6 ft   (c) 8 ft   (d) 10 ft

10. _____ must not be installed beneath openings through which materials may be moved, such as openings in farm and commercial buildings, and must not be installed where they will obstruct entrance to these building openings.

    (a) Overcurrent protection devices   (b) Overhead branch-circuit and feeder conductors
    (c) Grounding conductors   (d) Wiring systems

11. Raceways on exterior surfaces of buildings or other structures must be arranged to drain, and in _____ locations must be raintight.

    (a) damp   (b) wet   (c) dry   (d) all of these

12. Vegetation such as trees must not be used for support of _____.

    (a) overhead conductor spans   (b) surface wiring methods   (c) luminaires   (d) electric equipment

13. A building or structure must be supplied by a maximum of _____ feeder(s) or branch circuit(s).

    (a) one   (b) two   (c) three   (d) as many as desired

14. More than one feeder or branch circuit is permitted to supply a single building or other structure sufficiently large to require two or more supplies if permitted by _____.

    (a) architects   (b) special permission   (c) written authorization   (d) master electricians

15. The building disconnecting means must be installed at a(n) _____ location.

    (a) accessible   (b) readily accessible   (c) outdoor   (d) indoor

16. The disconnecting means is not required to be located at the building or structure where documented safe switching procedures are established and maintained, and where the installation is monitored by _____ persons.

    (a) maintenance   (b) management   (c) service   (d) qualified

17. •There must be no more than _____ disconnects installed for each electric supply.

    (a) two   (b) four   (c) six   (d) none of these

18. The two to six disconnects as permitted by 225.33 must be _____. Each disconnect must be marked to indicate the load served.

    (a) the same size   (b) grouped   (c) in the same enclosure   (d) none of these

19. The one or more additional disconnecting means for fire pumps or for emergency, legally required standby or optional standby systems as permitted by 225.30, must be installed sufficiently remote from the one to six disconnecting means for normal supply to minimize the possibility of _____ interruption of supply.

    (a) accidental   (b) intermittent   (c) simultaneous   (d) prolonged

(• Indicates that 75% or fewer exam takers get the question correct)

20. In a multiple-occupancy building, each occupant must have access to his or her own _____.

    (a) disconnecting means   (b) building drops   (c) building-entrance assembly   (d) lateral conductors

21. When the building disconnecting means is a power-operated switch or circuit breaker, it must be able to be opened by hand in the event of a _____.

    (a) ground fault   (b) short circuit   (c) power surge   (d) power-supply failure

22. •The building or structure disconnecting means must plainly indicate whether it is in the _____ position.

    (a) open or closed   (b) correct   (c) up or down   (d) none of these

(• Indicates that 75% or fewer exam takers get the question correct)

Article 230 Services

Article Overview

This article covers the installation requirements for service conductors and equipment. The requirements for service conductors differ from those for other conductors. For one thing, service conductors for one structure cannot pass through the interior of another structure [230.3], and you apply different rules depending on whether a service conductor is inside or outside a structure. When are they "outside" as opposed to "inside?" The answer may seem obvious, but isn't.

It's usually good to start a service installation by deciding which conductors actually are parts of the service. What you decide here will determine how you do the rest of the job. To identify a service conductor, you must first determine whether you're dealing with a service (line side) or a premises (load side) distribution point.

Let's review the following definitions in Article 100 to understand when the requirements of Article 230 apply:

- Service Point—The point of connection between the facilities of the serving utility and the premises wiring.

- Service Conductors—The conductors from the service point to the service disconnecting means (service equipment, not meter). Service-entrance conductors may be either overhead (service drop) or underground (service lateral).

- Service Equipment—The necessary equipment, usually consisting of circuit breakers or switches and fuses and their accessories, connected to the load end of service conductors in a building or other structure (or an otherwise designated area), and intended to constitute the main control and cutoff of the electricity supply. Service equipment doesn't include the metering equipment, such as the meter and meter enclosure [230.66].

After reviewing these definitions, you should understand that service conductors originate at the serving utility (service point) and terminate on the line side of the service disconnecting means (service equipment). Conductors and equipment on the load side of service equipment are considered feeder conductors and must be installed in accordance with Articles 215 and 225. These include:

- Secondary conductors from customer-owned transformers.
- Conductors from generators, UPS systems, or photovoltaic systems.
- Conductors to remote buildings or structures.

Conductors supplied from a battery, uninterruptible power supply system, solar photovoltaic system, generator, transformer, or phase converters aren't considered service conductors; they are feeder conductors [Article 100 Feeder].

Questions

1. A building or structure must be supplied by a maximum of _____ service(s).

    (a) one   (b) two   (c) three   (d) as many as desired

2. Additional services must be permitted for a single building or other structure sufficiently large to make two or more services necessary if permitted by _____.

    (a) architects   (b) special permission   (c) written authorization   (d) master electricians

(• Indicates that 75% or fewer exam takers get the question correct)

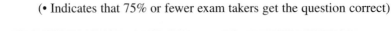

3. Additional services are permitted for different voltages, frequencies, or phases, or for different uses such as for _____.

   (a) gymnasiums (b) different rate schedules
   (c) flea markets (d) special entertainment events

4. Where a building or structure is supplied by more than one service, or a combination of branch circuits, feeders, and services, a permanent plaque or directory must be installed at each service disconnect location denoting all other _____ supplying that building or structure and the area served by each.

   (a) services (b) feeders (c) branch circuits (d) all of these

5. Service conductors supplying a building or other structure must not _____ of another building or other structure.

   (a) be installed on the exterior walls (b) pass through the interior (c) a and b (d) none of these

6. •Conductors other than service conductors must not be installed in the same _____.

   (a) service raceway (b) service cable (c) enclosure (d) a or b

7. Where a service raceway enters a building or structure from a(n) _____ it must be sealed in accordance with 300.5(G).

   (a) transformer vault (b) underground distribution system
   (c) cable tray (d) overhead rack

8. Overhead-service conductors to a building must maintain a vertical clearance of final spans above, or within, _____ measured horizontally from the platforms, projections, or surfaces from which they might be reached.

   (a) 3 ft (b) 6 ft (c) 8 ft (d) 10 ft

9. _____ must not be installed beneath openings through which materials may be moved, such as openings in farm and commercial buildings, and must not be installed where they will obstruct entrance to these building openings.

   (a) Overcurrent protection devices (b) Overhead-service conductors
   (c) Grounding conductors (d) Wiring systems

10. Overhead service conductors can be supported to hardwood trees.

    (a) True (b) False

11. Service-drop conductors must have _____.

    (a) sufficient ampacity to carry the current for the load (b) adequate mechanical strength
    (c) a or b (d) a and b

12. The minimum size service-drop conductor permitted by the *Code* is _____ AWG copper or _____ AWG aluminum or copper-clad aluminum.

    (a) 8, 6 (b) 6, 8 (c) 6, 6 (d) 8, 8

13. Service-drop conductors must have a minimum of _____ vertical clearance from final grade over residential property and driveways, as well as those commercial areas not subject to truck traffic where the voltage is limited to 300 volts-to-ground.

    (a) 10 ft (b) 12 ft (c) 15 ft (d) 18 ft

14. The minimum clearance for service drops not exceeding 600V that pass over commercial areas subject to truck traffic is _____.

    (a) 10 ft (b) 12 ft (c) 15 ft (d) 18 ft

(• Indicates that 75% or fewer exam takers get the question correct)

15. •Overhead service-drop conductors must have a horizontal clearance of _____ from a pool.

    (a) 6 ft　　(b) 10 ft　　(c) 8 ft　　(d) 4 ft

16. The minimum point of attachment of the service-drop conductors to a building must in no case be less than _____ above finished grade.

    (a) 8 ft　　(b) 10 ft　　(c) 12 ft　　(d) 15 ft

17. Service-lateral conductors are required to be insulated except for the grounded neutral conductor when it is _____.

    (a) bare copper used in a raceway
    (b) bare copper and part of a cable assembly that is identified for underground use
    (c) copper-clad aluminum
    (d) a or b

18. Underground copper service conductors must not be smaller than _____ AWG copper.

    (a) 3　　(b) 4　　(c) 6　　(d) 8

19. Service lateral conductors that supply power to limited loads of a single branch circuit must not be smaller than _____.

    (a) 4 AWG copper　　(b) 8 AWG aluminum　　(c) 12 AWG copper　　(d) none of these

20. When two to six service disconnecting means in separate enclosures are grouped at one location and supply separate loads from one service drop or lateral, _____ set(s) of service-entrance conductors are permitted to supply each or several such service equipment enclosures.

    (a) one　　(b) two　　(c) three　　(d) four

21. Service conductors must be sized at no less than _____ percent of the continuous load, plus 100 percent of the noncontinuous load.

    (a) 100　　(b) 115　　(c) 125　　(d) 150

22. Wiring methods permitted for service conductors include _____.

    (a) rigid metal conduit　　(b) electrical metallic tubing　　(c) rigid nonmetallic conduit　　(d) all of these

23. Cable tray systems are permitted to support service-entrance conductors. Cable trays used to support service-entrance conductors can contain only service-entrance conductors _____.

    (a) unless a solid fixed barrier separates the service-entrance conductors
    (b) only for under 300 volts
    (c) only in industrial locations
    (d) only for over 600 volts

24. •Service-entrance conductors must not be spliced or tapped.

    (a) True　　(b) False

25. Service-entrance conductors can be spliced or tapped by clamped or bolted connections at any time as long as _____.

    (a) the free ends of conductors are covered with an insulation that is equivalent to that of the conductors or with an insulating device identified for the purpose
    (b) wire connectors or other splicing means installed on conductors that are buried in the earth are listed for direct burial
    (c) no splice is made in a raceway
    (d) all of these

(• Indicates that 75% or fewer exam takers get the question correct)

26. Service cables, where subject to physical damage, must be protected.

    (a) True (b) False

27. Service cables that are subject to physical damage must be protected by _____.

    (a) rigid metal conduit
    (b) intermediate metal conduit
    (c) schedule 80 rigid nonmetallic conduit
    (d) any of these

28. Service cables must be equipped with a raintight _____.

    (a) raceway (b) service head (c) cover (d) all of these

29. Type SE cable is permitted to be formed in a _____ and taped with self-sealing weather-resistant thermoplastic.

    (a) loop (b) circle (c) gooseneck (d) none of these

30. Service heads must be located _____.

    (a) above the point of attachment
    (b) below the point of attachment
    (c) even with the point of attachment
    (d) none of these

31. To prevent water from entering service equipment, service-entrance conductors must _____.

    (a) be connected to service-drop conductors below the level of the service head
    (b) have drip loops formed on the service-entrance conductors
    (c) a or b
    (d) a and b

32. Service-drop conductors and service-entrance conductors must be arranged so that _____ will not enter the service raceway or equipment.

    (a) dust (b) vapor (c) water (d) none of these

33. On a 3Ø, 4-wire, delta-connected service where the midpoint of one phase winding is grounded, the service conductor having the higher-phase voltage-to-ground must be durably and permanently marked by an outer finish that is _____ in color, or by other effective means, at each termination or junction point.

    (a) orange (b) red (c) blue (d) any of these

34. The service disconnecting means must be installed at a(n) _____ location.

    (a) dry (b) readily accessible (c) outdoor (d) indoor

35. Service disconnecting means must not be installed in bathrooms.

    (a) True (b) False

36. Each service disconnecting means must be permanently marked to identify it as a service disconnecting means.

    (a) True (b) False

37. Each service disconnecting means must be permanently _____ to identify it as a service disconnect.

    (a) identified (b) positioned (c) marked (d) none of these

(• Indicates that 75% or fewer exam takers get the question correct)

38. There must be no more than _____ disconnects installed for each service or for each set of service-entrance conductors as permitted in 230.2 and 230.40.

    (a) two   (b) four   (c) six   (d) none of these

39. Disconnecting means used solely for power monitoring equipment, transient voltage surge suppressors, or the control circuit of the ground-fault protection system or power-operable service disconnecting means, installed as part of the listed equipment, are not considered a service disconnecting means.

    (a) True   (b) False

40. •When the service contains two to six service disconnecting means, they must be _____.

    (a) the same size   (b) grouped at one location   (c) in the same enclosure   (d) none of these

41. The additional service disconnecting means permitted by 230.2 for fire pumps or for emergency, legally required standby, or optional standby services, must be installed remote from the one to six service disconnecting means for normal service to minimize the possibility of _____ interruption of supply.

    (a) accidental   (b) intermittent   (c) simultaneous   (d) prolonged

42. When the service disconnecting means is a power-operated switch or circuit breaker, it must be able to be opened by hand in the event of a _____.

    (a) ground fault   (b) short circuit   (c) power surge   (d) power-supply failure

43. •The service disconnecting means must plainly indicate whether it is in the _____ position.

    (a) open or closed   (b) tripped   (c) up or down   (d) correct

44. When the service disconnecting means consists of more than one switch or circuit breaker, the combined ratings of all the switches or circuit breakers used _____ than the rating required by 230.79.

    (a) must be less   (b) must not be less   (c) must be more   (d) none of these

45. The service conductors must be connected to the service disconnecting means by _____ or other approved means.

    (a) pressure connectors (b) clamps   (c) solder   (d) a or b

46. Meter disconnect switches that have a short-circuit current rating equal to or greater than the available short-circuit current are permitted ahead of the service-disconnecting means.

    (a) True   (b) False

47. _____ for power-operable service disconnects can be connected on the supply side of the service disconnecting means, if suitable overcurrent protection and disconnecting means are provided.

    (a) Control circuits   (b) Distribution panels (c) Grounding conductors (d) none of these

48. Each _____ service conductor must have overload protection.

    (a) overhead   (b) underground   (c) ungrounded   (d) none of these

49. Circuits used only for the operation of fire alarms, other protective signaling systems, or the supply to fire pump equipment are permitted to be connected on the _____ of the service overcurrent protection device where separately provided with overcurrent protection.

    (a) base   (b) load side   (c) supply side   (d) top

(• Indicates that 75% or fewer exam takers get the question correct)

Article 240 Overcurrent Protection

Article Overview

This article provides the requirements for selecting and installing overcurrent protection devices. A review of the basic concept of overcurrent protection will help you avoid confusion as we move forward. Overcurrent exists when current exceeds the rating of conductors or equipment. This can be due to overload, short circuit, or ground fault.

Overload. An overload is a condition where equipment or conductors carry current exceeding their rated ampacity. An example of an overload is plugging two 12.5A (1,500W) hair dryers into a 20A branch circuit.

Ground Fault. A ground fault is an unintentional, electrically conducting connection between an ungrounded conductor of an electrical circuit and the normal noncurrent-carrying conductors, metallic enclosures, metallic raceways, metallic equipment, or earth. During the period of a ground fault, dangerous voltages and larger than normal currents exist.

Short Circuit. A short circuit is the unintentional electrical connection between any two normally current-carrying conductors of an electrical circuit, either line-to-line or line-to-neutral.

Overcurrent protection devices protect conductors and equipment. But it's important to note that they protect conductors differently than equipment.

An overcurrent protection device protects a *circuit* by opening when current reaches a value that would cause an excessive temperature rise in conductors. The overcurrent protection device's interrupting rating must be sufficient for the maximum possible fault current available on the line-side terminals of the equipment [110.9]. You'll find the standard ratings for fuses and *fixed-trip* circuit breakers in 240.6. Using a water analogy, current rises like water in a tank—at a certain level, the overcurrent protection device shuts off the faucet. Think in terms of normal operating conditions that just get too far out of normal range.

An overcurrent protection device protects *equipment* by opening when it detects a short circuit or ground fault. Every piece of electrical equipment must have a short-circuit current rating that permits the overcurrent protection devices (for that equipment) to clear short circuits or ground faults without extensive damage to the electrical components of the circuit [110.10]. Using a water analogy, a pipe bursts—creating a sudden rise in level—and the overcurrent protection device shuts off the supply to the pipe. Short circuits and ground faults aren't normal operating conditions. Thus, the overcurrent protection devices for equipment will have different characteristics than overcurrent protection devices for conductors.

Questions

1. Overcurrent protection for conductors and equipment is designed to _____ the circuit if the current reaches a value that will cause an excessive or dangerous temperature in conductors or conductor insulation.

    (a) open   (b) close   (c) monitor   (d) record

2. A device that, when interrupting currents in its current-limiting range, will reduce the current flowing in the faulted circuit to a magnitude substantially less than that obtainable in the same circuit if the device were replaced with a solid conductor having comparable impedance, is defined as a(n) _____ protective device.

    (a) short-circuit   (b) overload   (c) ground-fault   (d) current-limiting

(• Indicates that 75% or fewer exam takers get the question correct)

3. Conductor overload protection is not required where the interruption of the _____ would create a hazard, such as in a material-handling magnet circuit or fire-pump circuit. However, short-circuit protection is required.

   (a) circuit   (b) line   (c) phase   (d) system

4. Unless specifically permitted in 240.4(E) through 240.4(G), the overcurrent protection must not exceed _____ after any correction factors for ambient temperature and the number of conductors has been applied.

   (a) 15A for 14 AWG copper   (b) 20A for 12 AWG copper   (c) 30A for 10 AWG copper   (d) all of these

5. 240.4(E) allows tap conductors to be protected against overcurrent in accordance with other Code sections that deal with the specific situation outside Article 240.

   (a) True   (b) False

6. Flexible cords approved for and used with a specific listed appliance or portable lamp are considered to be protected when _____.

   (a) not more than 6 ft in length   (b) 20 AWG and larger
   (c) applied within the listing requirements   (d) 16 AWG and larger

7. Where flexible cord is used in listed extension cord sets, the conductors are considered protected against overcurrent when used within _____.

   (a) indoor installations   (b) non-hazardous locations
   (c) the extension cord's listing requirements   (d) 50 ft of the branch-circuit panelboard

8. Which of the following is not a standard size for fuses or inverse-time circuit breakers?

   (a) 45A   (b) 70A   (c) 75A   (d) 80A

9. The standard ampere ratings for fuses and inverse-time circuit breakers are listed in 240.6(a). Additional standard ratings for fuses include _____.

   (a) 1A   (b) 6A   (c) 601A   (d) all of these

10. Supplementary overcurrent protection _____.

    (a) must not be used in luminaires.
    (b) may be used as a substitute for a branch-circuit overcurrent protection device.
    (c) may be used to protect internal circuits of equipment.
    (d) must be readily accessible.

11. Supplementary overcurrent devices used in luminaires or appliances are not required to be readily accessible.

    (a) True   (b) False

12. Except where limited by 210.4(B), individual single-pole circuit breakers, with or without approved handle ties, are permitted as the protection for each ungrounded conductor of multiwire branch circuits that serve only 1Ø line-to-neutral loads.

    (a) True   (b) False

13. Single-pole breakers with identified handle ties can be used to protect each ungrounded conductor for line-to-line connected loads.

    (a) True   (b) False

(• Indicates that 75% or fewer exam takers get the question correct)

14. No tap conductor can supply another tap conductor.

    (a) True   (b) False

15. A feeder tap of 10 ft or less can be made without overcurrent protection at the tap when the rating of the overcurrent device on the line side of the tap conductors does not exceed _____ times the ampacity of the tap conductor.

    (a) 10   (b) 5   (c) 125   (d) 25

16. Overcurrent protection for tap conductors not over 25 ft is not required at the point where the conductors receive their supply providing the _____.

    (a) ampacity of the tap conductors is not less than one-third the rating of the overcurrent device protecting the feeder conductors being tapped
    (b) tap conductors terminate in a single circuit breaker or set of fuses that limit the load to the ampacity of the tap conductors
    (c) tap conductors are suitably protected from physical damage
    (d) all of these

17. One of the requirements that permit conductors supplying a transformer to be tapped, without overcurrent protection at the tap, is that the conductors supplied by the _____ of a transformer must have an ampacity, when multiplied by the ratio of the primary-to-secondary voltage, of at least one-third the rating of the overcurrent device protecting the feeder conductors.

    (a) primary   (b) secondary   (c) tertiary   (d) none of these

18. The maximum length of an unprotected feeder tap conductor in a high-bay manufacturing building over 35 ft high is _____.

    (a) 15 ft   (b) 20 ft   (c) 50 ft   (d) 100 ft

19. The "next size up protection rule" of 240.4(B) is permitted for transformer secondary tap conductors.

    (a) True   (b) False

20. For industrial installations only, a tap can be made without overcurrent protection when the transformer secondary conductors have a total length of not more than _____. The tap conductors must have an ampacity not less than the secondary current rating of the transformer and the sum of the ratings of the overcurrent devices.

    (a) 8 ft   (b) 25 ft   (c) 35 ft   (d) 75 ft

21. Circuit breakers and fuses must be readily accessible and they must be installed so the center of the grip of the operating handle of the fuse switch or circuit breaker, when in its highest position, isn't more than _____ above the floor or working platform.

    (a) 6 ft 7 in.   (b) 2 ft   (c) 5 ft   (d) 4 ft 6 in.

22. Overcurrent protection devices must be _____.

    (a) accessible (as applied to wiring methods)   (b) accessible (as applied to equipment)
    (c) readily accessible   (d) inaccessible to unauthorized personnel

23. Branch-circuit overcurrent protection devices are not required to be accessible to occupants of guest rooms of hotels and motels if maintenance is provided in a facility that is under continuous building management.

    (a) True   (b) False

(• Indicates that 75% or fewer exam takers get the question correct)

24. Overcurrent protection devices are not permitted to be located _____.

   (a) where exposed to physical damage
   (b) near easily ignitable materials, such as in clothes closets
   (c) in bathrooms of dwelling units
   (d) all of these

25. Enclosures for overcurrent protection devices must be mounted in a _____ position unless that is shown to be impracticable.

   (a) vertical   (b) horizontal   (c) vertical or horizontal   (d) there are no requirements

26. Plug fuses of 15A or less must be identified by a(n) _____ configuration of the window, cap, or other prominent part to distinguish them from fuses of higher ampere ratings.

   (a) octagonal   (b) rectangular   (c) hexagonal   (d) triangular

27. Plug fuses of the Edison-base type have a maximum rating of _____.

   (a) 20A   (b) 30A   (c) 40A   (d) 50A

28. Plug fuses of the Edison-base type must be used _____.

   (a) where overfusing is necessary
   (b) only as replacement in existing installations
   (c) as a replacement for Type S fuses
   (d) only for 50A and above

29. Fuseholders of the Edison-base type must be installed only where they are made to accept _____ fuses by the use of adapters.

   (a) Edison-base   (b) medium-base   (c) heavy-duty base   (d) Type S

30. •Which of the following statements about Type S fuses is (are) true?

   (a) Adapters must fit Edison-base fuseholders.
   (b) Adapters are designed to be easily removed.
   (c) Type S fuses must be classified as not over 125V and 30A.
   (d) a and c

31. Type _____ fuse adapters must be designed so that once inserted in a fuseholder they cannot be removed.

   (a) A   (b) E   (c) S   (d) P

32. Type S fuses, fuseholders, and adapters are required to be designed so that _____ would be difficult.

   (a) installation   (b) tampering   (c) shunting   (d) b or c

33. Cartridge fuses and fuseholders of the 300V type are not permitted on circuits exceeding 300V _____.

   (a) between conductors   (b) to ground   (c) or less   (d) a or c

34. Fuseholders for cartridge fuses must be so designed that it is difficult to put a fuse of any given class into a fuseholder that is designed for a _____ lower or a _____ higher than that of the class to which the fuse belongs.

   (a) voltage, wattage   (b) wattage, voltage   (c) voltage, current   (d) current, voltage

35. Fuses are required to be marked with _____.

   (a) ampere and voltage rating
   (b) interrupting rating where other than 10,000A
   (c) the name or trademark of the manufacturer
   (d) all of these

36. Cartridge fuses and fuseholders must be classified according to _____ ranges.

   (a) voltage   (b) amperage   (c) voltage or amperage   (d) voltage and amperage

(• Indicates that 75% or fewer exam takers get the question correct)

37. Circuit breakers must be capable of being closed and opened by manual operation. Their normal method of operation by other means, such as electrical or pneumatic must be permitted if means for _____ operation are also provided.

    (a) automated   (b) timed   (c) manual   (d) shunt trip

38. Circuit breakers must clearly indicate whether they are in the open "off" or closed "on" position. Where the circuit breaker handles are operated vertically the "up" position of the handle must be the _____.

    (a) "on" position   (b) "off" position   (c) tripped position   (d) any of these

39. Circuit breakers must be marked with their _____ rating in a manner that will be durable and visible after installation.

    (a) voltage   (b) ampere   (c) type   (d) all of these

40. Circuit breakers must be marked with their ampere rating in a manner that will be durable and visible after installation. Such marking can be made visible by removal of a _____.

    (a) trim   (b) cover   (c) box   (d) a or b

41. Circuit breakers rated at _____ amperes or less and _____ volts or less must have the ampere rating molded, stamped, etched, or similarly marked into their handles or escutcheon areas.

    (a) 100, 600   (b) 600, 100   (c) 1,000, 6,000 (d) 6,000, 1,000

42. Circuit breakers having an interrupting current rating of other than _____ must have their interrupting rating marked on the circuit breaker.

    (a) 50,000A   (b) 10,000A   (c) 15,000A   (d) 5,000A

43. Circuit breakers used as switches in 120V or 277V fluorescent-lighting circuits must be listed and marked _____.

    (a) UL   (b) SWD or HID   (c) Amps   (d) VA

44. Circuit breakers used to switch high-intensity discharge lighting circuits must be listed and marked as _____.

    (a) SWD   (b) HID   (c) a or b   (d) a and b

45. A circuit breaker with a straight voltage rating (240V or 480V) can be used on a circuit where the nominal voltage between any two conductors does not exceed the circuit breaker's voltage rating.

    (a) True   (b) False

46. A circuit breaker with a slash rating (120/240V or 277/480V) can be used on a solidly-grounded circuit where the nominal voltage of any conductor to _____ does not exceed the lower of the two values, and the nominal voltage between any two conductors does not exceed the higher value.

    (a) another conductor (b) an enclosure (c) earth   (d) ground

(• Indicates that 75% or fewer exam takers get the question correct)

Article 250 Grounding and Bonding

Article Overview

The purpose of the *National Electrical Code* is the practical safeguarding of persons and property from hazards arising from the use of electricity [90.1(A)]. In addition, the *NEC* contains provisions that are considered necessary for safety. Compliance with the *NEC*, combined with proper maintenance, must result in an installation that is essentially free from hazard [90.1(B)].

No other article can match Article 250 for misapplication, violation, and misinterpretation. People often insist on completing installations in a manner that results in violations of this article. For example, many industrial equipment manuals require violating 250.4(A)(5) as a condition of warranty. The manuals insist on installing an "isolated grounding electrode," which is an electrode without a low-impedance fault-current path back to the electrical supply source, typically the X0 terminal of a transformer, other than through the earth itself. That means the ground-fault current return path to the electrical supply source, typically the X0 terminal (utility transformer), is on the order of several ohms rather than the fraction of an ohm that the typical *NEC*-compliant installation would provide.

If you apply basic physics and basic electrical theory, you can clearly see Article 250 is right and equipment manuals that require isolated grounding are wrong, and other references agree. IEEE-142 and *Soares Book on Grounding* use the same physics and electrical theory as Article 250. This article isn't a "preferred design specification." As with the rest of the *NEC*, it serves the purpose stated in Article 90 to be sure the installation is, and remains, SAFE!

Article 250 covers the requirements for providing paths to divert high voltage to the earth, requirements for the low-impedance fault-current path to facilitate the operation of overcurrent protection devices, and how to remove dangerous voltage potentials between conductive parts of building components and electrical systems.

Over the past two *Code* cycles, this article was extensively revised to make it better organized and easier to implement. It's arranged in a logical manner, so it's a good idea to just read through Article 250 to get a big picture view—after you review the definitions. Then study the article closely so you understand the details. The illustrations in the *Understanding the NEC, Volume 1* textbook will help you understand the key points.

> **Author's Comment:** When the *NEC* uses the word "ground" or "grounding" where the intent is the connection to the earth, this workbook will add "(earth)." If the *NEC* used the word "ground" or "grounding" where the intent is bonding metal parts to the supply source, I will add "(bonding)" to the text. It's unfortunate that the word "grounding" is used interchangeably for grounding and bonding, which are two different things.

Questions

1. A ground-fault current path is an electrically conductive path from the point of a line-to-case fault extending to the _____.

    (a) ground   (b) earth   (c) electrical supply source   (d) none of these

2. An effective ground-fault current path is an intentionally constructed low-impedance path designed and intended to carry fault current from the point of a line-to-case fault on a wiring system to _____.

    (a) ground   (b) earth   (c) the electrical supply source   (d) none of these

3. An effective ground-fault current path is created when all electrically conductive materials that are likely to be energized are bonded together and to the _____.

    (a) ground   (b) earth   (c) electrical supply source   (d) none of these

(• Indicates that 75% or fewer exam takers get the question correct)

4. Electrical systems that are grounded must be connected to earth in a manner that will _____.

   (a) limit voltages due to lightning, line surges, or unintentional contact with higher voltage lines
   (b) stabilize the voltage-to-ground during normal operation
   (c) facilitate overcurrent protection device operation in case of ground faults
   (d) a and b

5. Electrical systems that are grounded, including transformers and generators, must be connected to the _____ for the purpose of limiting the voltage imposed by lightning, line surges, or unintentional contact with higher voltage lines.

   (a) ground   (b) earth   (c) electrical supply source   (d) none of these

6. Electrical systems are grounded to the _____ to stabilize the system voltage.

   (a) ground   (b) earth   (c) electrical supply source   (d) none of these

7. For grounded systems, noncurrent-carrying conductive materials enclosing electrical conductors or equipment, or forming part of such equipment, must be connected to earth so as to limit the voltage-to-ground on these materials.

   (a) True   (b) False

8. For grounded systems, the metal parts of electrical equipment in a building or structure must be connected to the _____ for the purpose of limiting the voltage to ground on these materials.

   (a) ground   (b) earth   (c) electrical supply source   (d) none of these

9. For grounded systems, noncurrent-carrying conductive materials enclosing electrical conductors or equipment, or forming part of such equipment, must be connected together and to the _____ in a manner that establishes an effective ground-fault current path.

   (a) ground   (b) earth   (c) electrical supply source   (d) none of these

10. For grounded systems, the electrical equipment and wiring, and other electrically conductive material likely to become energized, are installed in a manner that creates a permanent, low-impedance circuit capable of safely carrying the maximum ground-fault current likely to be imposed on it from where a ground fault may occur to the _____.

    (a) ground   (b) earth   (c) electrical supply source   (d) none of these

11. For grounded systems, electrical equipment and wiring, and other electrically conductive material likely to become energized, must be installed in a manner that creates a _____ from any point on the wiring system where a ground fault may occur to the electrical supply source.

    (a) permanent path
    (b) low-impedance path
    (c) path capable of safely carrying the ground-fault current likely to be imposed on it
    (d) all of these

12. For grounded systems, the earth can be considered as an effective ground-fault current path.

    (a) True   (b) False

13. For ungrounded systems, noncurrent-carrying conductive materials enclosing electrical conductors or equipment, or forming part of such equipment, must be connected to earth in a manner that will limit the voltage imposed by lightning or unintentional contact with higher-voltage lines.

    (a) True   (b) False

(• Indicates that 75% or fewer exam takers get the question correct)

14. The grounding of electrical systems, circuit conductors, surge arresters, and conductive noncurrent-carrying materials and equipment must be installed and arranged in a manner that will prevent objectionable current over the grounding conductors or grounding paths.

    (a) True   (b) False

15. Temporary current flowing on the effective ground-fault current path during a ground fault condition is considered by the *Code* to be objectionable current.

    (a) True   (b) False

16. •Currents that introduce noise or data errors in electronic equipment are considered objectionable currents.

    (a) True   (b) False

17. Grounding and bonding conductors cannot be connected by _____.

    (a) pressure connections   (b) solder   (c) lugs   (d) approved clamps

18. Sheet-metal screws can be used to connect grounding (or bonding) conductors or connection devices to enclosures.

    (a) True   (b) False

19. Grounding electrode conductor fittings must be protected from physical damage by being enclosed in _____ where there may be a possibility of physical damage.

    (a) metal   (b) wood   (c) the equivalent of a or b   (d) none of these

20. _____ on equipment to be grounded must be removed from contact surfaces to ensure good electrical continuity.

    (a) Paint   (b) Lacquer   (c) Enamel   (d) any of these

21. AC circuits of less than 50 volts must be grounded if supplied by a transformer whose supply system exceeds 150 volts-to-ground.

    (a) True   (b) False

22. AC systems of 50 to 1,000 volts that supply premises wiring systems must be grounded where the system can be grounded so that the maximum voltage-to-ground on the ungrounded conductors does not exceed _____.

    (a) 1,000 volts   (b) 300 volts   (c) 150 volts   (d) 50 volts

23. AC systems of 50 to 1,000 volts that supply premises wiring systems must be grounded where supplied by a 3Ø, 4-wire, wye connected system.

    (a) True   (b) False

24. AC systems of 50 to 1,000 volts that supply premises wiring systems must be grounded where supplied by a 3Ø, 4-wire, delta connected system.

    (a) True   (b) False

25. An alternate ac power source such as an onsite generator is not a separately derived system if the _____ is solidly interconnected to a service-supplied system neutral.

    (a) ignition system   (b) fuel cell   (c) neutral   (d) line conductor

(• Indicates that 75% or fewer exam takers get the question correct)

26. •When grounding service-supplied alternating-current systems, the grounding electrode conductor must be connected (bonded) to the grounded service conductor (neutral) at _____.

    (a) the load end of the service drop
    (b) the meter equipment
    (c) the service disconnect
    (d) any of these

27. The grounding electrode conductor at the service is permitted to terminate on an equipment grounding terminal bar if a (main) bonding jumper is installed between the grounded neutral conductor bus and the equipment grounding terminal.

    (a) True (b) False

28. A grounding connection must not be made to any grounded circuit conductor on the _____ side of the service disconnecting means except as permitted for separately derived systems or separate buildings.

    (a) supply (b) power (c) line (d) load

29. Where an ac system operating at less than 1,000V is grounded at any point, the _____ conductors must be run to each service disconnecting means and must be bonded to each disconnect enclosure.

    (a) ungrounded (b) grounded (c) grounding (d) none of these

30. The grounded neutral conductor brought to service equipment must be routed with the phase conductors and must not be smaller than specified in Table _____ when the service-entrance conductors are not larger than 1,100 kcmil copper.

    (a) 250.66 (b) 250.122 (c) 310.16 (d) 430.52

31. When service-entrance conductors exceed 1,100 kcmil for copper, the required grounded neutral conductor for the service must be sized not less than _____ percent of the area of the largest ungrounded service-entrance (phase) conductor.

    (a) 15 (b) 19 (c) 12½ (d) 25

32. Where the service-entrance phase conductors are installed in parallel, the size of the grounded neutral conductor in each raceway must be based on the size of the ungrounded service-entrance conductor in the raceway, but not smaller than _____ AWG.

    (a) 6 (b) 1 (c) 1/0 (d) none of these

33. For a grounded system, an unspliced _____ must be used to connect the equipment grounding conductor(s) and the service disconnect enclosure to the grounded neutral conductor of the system within the enclosure for each service disconnect.

    (a) grounding electrode (b) main bonding jumper
    (c) bus bar only (d) insulated copper conductor only

34. A main bonding jumper must be a _____ or similar conductor.

    (a) wire (b) bus (c) screw (d) any of these

35. Where a main bonding jumper is a screw only, the screw must be identified with _____ that must be visible with the screw installed.

    (a) a silver or white finish (b) an etched ground symbol (c) a green tag (d) a green finish

36. The grounding electrode conductor for a single separately derived system must connect the grounded neutral conductor of the derived system to the grounding electrode.

    (a) True (b) False

(• Indicates that 75% or fewer exam takers get the question correct)

37. For a single separately derived system, the grounding electrode conductor connects the grounding electrode to the grounded neutral conductor of the derived system at the same point on the separately derived system where the _____ is installed.

    (a) metering equipment   (b) transfer switch   (c) bonding jumper   (d) largest circuit breaker

38. Grounding electrode taps from a separately derived system to a common grounding electrode conductor are permitted when a building or structure has multiple separately derived systems.

    (a) True   (b) False

39. The grounding electrode for a separately derived system must be as near as practicable to, and preferably in the same area as, the grounding electrode conductor connection to the system. The grounding electrode must be the nearest one of the following:

    (a) An effectively grounded metal member of the building structure.
    (b) An effectively grounded metal water pipe, but only if it's within 5 ft from the point of entrance into the building.
    (c) Any metal structure that is effectively grounded.
    (d) a or b

40. Where a grounded neutral conductor is installed and the neutral-to-case bond is not at the source of the separately derived system, the grounded neutral conductor must be routed with the derived phase conductors and must not be smaller than the required grounding electrode conductor specified in Table 250.66, but must not be required to be larger than the largest ungrounded derived phase conductor.

    (a) True   (b) False

41. A grounding electrode is required if a building or structure is supplied by a feeder or by more than one branch circuit.

    (a) True   (b) False

42. A grounding electrode at a separate building or structure is required where one multiwire branch circuit serves the building or structure.

    (a) True   (b) False

43. When supplying a grounded system at a separate building or structure, if the equipment grounding conductor is run with the supply conductors and connected to the building disconnecting means, there must be no connection made between the grounded neutral conductor and the equipment grounding conductor at the separate building.

    (a) True   (b) False

44. When supplying a grounded system at a separate building or structure, if the equipment grounding conductor is not run with the supply conductors and there are no continuous metallic paths bonded to the grounding system in both buildings involved, and ground fault protection of equipment has not been installed on the common ac service, then the grounded circuit conductor must be connected to the building disconnecting means and to the grounding electrode at the separate building.

    (a) True   (b) False

45. The size of the grounding electrode conductor for a building or structure supplied by a feeder cannot be smaller than that identified in _____ based on the largest ungrounded supply conductor.

    (a) 250.66   (b) 250.122   (c) Table 310.16   (d) not specified

(• Indicates that 75% or fewer exam takers get the question correct)

46. The frame of a portable generator is not required to be grounded and is not to be connected to a(n) _____ for a system supplied by cord and plug using receptacles mounted on the generator with the grounding terminals of the receptacles bonded to the generator frame.

    (a) grounding electrode
    (b) grounded neutral conductor
    (c) ungrounded conductor
    (d) equipment grounding conductor

47. The frame of a vehicle-mounted generator is not required to be connected to a(n) _____ for a system supplied by cord and plug using receptacles mounted on the vehicle or the generator when the grounding terminals of the receptacles are bonded to the generator frame and the generator frame is bonded to the vehicle frame.

    (a) grounding electrode
    (b) grounded neutral conductor
    (c) ungrounded conductor
    (d) equipment grounding conductor

48. Where none of the items in 250.52(A)(1) through (A)(6) are present for use as a grounding electrode, one or more of the following must be installed and used as the grounding electrode: _____.

    (a) a ground ring
    (b) rod and pipe electrodes or plate electrodes
    (c) local metal underground systems or structures
    (d) any of these

49. Concrete-encased electrodes of _____ are not required to be part of the grounding electrode system where the steel reinforcing bars or rods aren't accessible for use without disturbing the concrete.

    (a) hazardous locations
    (b) health care facilities
    (c) existing buildings or structures
    (d) agricultural buildings with equipotential planes

50. Interior metal water piping located more than _____ from the point of entrance to the building cannot be used as a part of the grounding electrode system, or as a conductor to interconnect electrodes that are part of the grounding electrode system.

    (a) 2 ft
    (b) 4 ft
    (c) 5 ft
    (d) 6 ft

51. The metal frame of a building where one of the four *Code*-prescribed methods of making an earth connection has been met may serve as part of the grounding electrode system.

    (a) True
    (b) False

52. A bare 4 AWG copper conductor installed near the bottom of a concrete foundation or footing that is in direct contact with the earth may be used as a grounding electrode when the conductor is at least _____ in length.

    (a) 25 ft
    (b) 15 ft
    (c) 10 ft
    (d) 20 ft

53. Grounding electrodes that are driven rods require a minimum of _____ in contact with the soil.

    (a) 10 ft
    (b) 8 ft
    (c) 6 ft
    (d) 12 ft

54. Electrodes of pipe or conduit must not be smaller than _____ and, where of iron or steel, must have the outer surface galvanized or otherwise metal-coated for corrosion protection.

    (a) ½ in.
    (b) ¾ in.
    (c) 1 in.
    (d) none of these

55. Grounding electrodes consisting of stainless-steel rods or nonferrous rods that are less than 5⁄8 in. in diameter must be listed and cannot be less than _____ in diameter.

    (a) ½ in.
    (b) ¾ in.
    (c) 1 in.
    (d) 1¼

(• Indicates that 75% or fewer exam takers get the question correct)

56. A metal underground water pipe must be supplemented by an additional electrode of a type specified in 250.52(A)(2) through (A)(7). Where the supplemental electrode is a rod, pipe, or plate electrode, that portion of the bonding jumper that is the sole connection to the supplemental grounding electrode is not required to be larger than _____ AWG copper wire.

   (a) 8   (b) 6   (c) 4   (d) 1

57. The upper end of the rod electrode must be _____ ground level unless the aboveground end and the grounding electrode conductor attachment are protected against physical damage.

   (a) above   (b) flush with (c) below   (d) b or c

58. Ground rod electrodes must be installed so that at least _____ of the length is in contact with the soil. Where rock bottom is encountered, the rod must be driven at an angle not to exceed 45 degrees.

   (a) 8 ft   (b) 5 ft   (c) one half   (d) 80 percent

59. When driving a ground rod electrode, if rock bottom is encountered, the rod must be driven at an angle not to exceed 45 degrees. Where rock bottom is encountered when driving at an angle up to 45 degrees, the electrode is permitted to be buried in a trench that is at least _____ deep.

   (a) 4 ft   (b) 30 in.   (c) 8 ft   (d) 18 in.

60. When driving a ground rod electrode, if rock bottom is encountered, the rod is allowed to be bent over in a trench and buried or shortened with a hack saw.

   (a) True   (b) False

61. For electrical equipment supplementary electrodes:

    (1) A bond to the grounding electrode system is not required.

    (2) The bonding jumper to the supplementary electrode can be any size.

    (3) The 25 ohm resistance requirement of 250.56 does not apply.

   (a) True   (b) False

62. The supplementary electrode allowed by the *Code* is different from a supplemental electrode, and is allowed to be connected to the equipment grounding conductors but cannot be used in place of an effective ground-fault current path for electrical equipment.

   (a) True   (b) False

63. •Where the resistance-to-ground of a single rod electrode exceeds 25 ohms, _____.

    (a) other means besides made electrodes must be used in order to provide grounding
    (b) at least one additional electrode must be added
    (c) no additional electrodes are required
    (d) the electrode can be omitted

64. When multiple ground rods are used for a grounding electrode, they must be separated not less than _____ apart.

   (a) 6 ft   (b) 8 ft   (c) 20 ft   (d) 12 ft

65. Two or more grounding electrodes that are effectively bonded together are considered as a single grounding electrode system in this sense.

   (a) True   (b) False

(• Indicates that 75% or fewer exam takers get the question correct)

66. Where separate services supply a building and are required to be connected to a grounding electrode, the same grounding electrode must be used. Two or more grounding electrodes that are _____ are considered as a single grounding electrode system in this sense.

    (a) effectively bonded together
    (b) spaced no more than 6 ft apart
    (c) a and b
    (d) none of these

67. Air terminal conductors or electrodes used for grounding air terminals _____ be used as the grounding electrodes required by 250.50 for grounding wiring systems and equipment.

    (a) must   (b) must not   (c) can   (d) any of these

68. The grounding electrode conductor must be made of which of the following materials?

    (a) Copper   (b) Aluminum   (c) Copper-clad aluminum   (d) any of these

69. Bare aluminum or copper-clad aluminum grounding conductors must not be used where in direct contact with masonry, the earth, or where subject to corrosive conditions. Where used outside, aluminum or copper-clad aluminum grounding electrode conductors must not be terminated within _____ of the earth.

    (a) 6 in.   (b) 12 in.   (c) 15 in.   (d) 18 in.

70. Grounding electrode conductors smaller than _____ must be in rigid metal conduit, intermediate metal conduit, rigid nonmetallic conduit, electrical metallic tubing, or cable armor.

    (a) 6 AWG   (b) 8 AWG   (c) 10 AWG   (d) 4 AWG

71. Grounding electrode conductors _____ and larger that are not subject to physical damage can be run exposed along the surface, if securely fastened to the construction.

    (a) 6 AWG   (b) 8 AWG   (c) 10 AWG   (d) 4 AWG

72. The grounding electrode conductor must be installed in one continuous length without a splice or joint, unless spliced _____.

    (a) by connecting to a busbar
    (b) by irreversible compression-type connectors listed as grounding and bonding
    (c) by the exothermic welding process.
    (d) any of these

73. •Metal enclosures for grounding electrode conductors must be electrically continuous, from the point of attachment to cabinets or equipment, to the grounding electrode.

    (a) True   (b) False

74. The grounding electrode conductor can be run to any convenient grounding electrode available in the grounding electrode system or to one or more grounding electrodes individually.

    (a) True   (b) False

75. •A service that contains 12 AWG service-entrance conductors, as permitted by 230.23(B) Ex., requires a grounding electrode conductor sized no less than _____.

    (a) 6 AWG   (b) 4 AWG   (c) 8 AWG   (d) 10 AWG

(• Indicates that 75% or fewer exam takers get the question correct)

76. •The largest size grounding electrode conductor required for any service is a _____ copper.

    (a) 6 AWG   (b) 1/0 AWG   (c) 3/0 AWG   (d) 250 kcmil

77. •What size copper grounding electrode conductor is required for a service that has three sets of 500 kcmil copper conductors per phase?

    (a) 1 AWG   (b) 1/0 AWG   (c) 2/0 AWG   (d) 3/0 AWG

78. In an ac system, the size of the grounding electrode conductor to a concrete-encased electrode is not required to be larger than _____ copper wire.

    (a) 4 AWG   (b) 6 AWG   (c) 8 AWG   (d) 10 AWG

79. The connection of the grounding electrode conductor to a buried grounding electrode (driven ground rod) must be made with a listed terminal device that is accessible.

    (a) True   (b) False

80. Grounding electrode conductor connections to a concrete-encased or buried grounding electrode are required to be readily accessible.

    (a) True   (b) False

81. An exothermic or irreversible compression connection to fireproofed structural metal is required to be accessible.

    (a) True   (b) False

82. The connection (attachment) of the grounding electrode conductor to a grounding electrode must _____.

    (a) be accessible
    (b) be made in a manner that will ensure a permanent and effective grounding path
    (c) a and b
    (d) none of these

83. When an underground metal water-piping system is used as a grounding electrode, effective bonding must be provided around insulated joints and around any equipment that is likely to be disconnected for repairs or replacement. Bonding conductors must be of _____ to permit removal of such equipment while retaining the integrity of the bond.

    (a) stranded wire   (b) flexible conduit   (c) sufficient length   (d) none of these

84. The grounding conductor connection to the grounding electrode must be made by _____.

    (a) listed lugs   (b) exothermic welding   (c) listed pressure connectors   (d) any of these

85. Metal enclosures and raceways for service conductors and equipment must be _____.

    (a) isolated   (b) insulated   (c) grounded   (d) gray

86. A metal elbow that is installed in an underground installation of rigid nonmetallic conduit and is isolated from possible contact by a minimum cover _____ to any part of the elbow, is not required to be grounded.

    (a) of 6 in.   (b) of 12 in.
    (c) of 18 in.   (d) as specified in Table 300.5

(• Indicates that 75% or fewer exam takers get the question correct)

87. Metal enclosures and raceways for other than service conductors must be grounded except as permitted by 250.112(I).

    (a) True   (b) False

88. Metal enclosures and raceways for conductors added to existing installations of _____, which do not provide an equipment ground are not required to be grounded if they are less than 25 ft long, they are free from probable contact with grounded conductive material, and are guarded against contact by persons.

    (a) nonmetallic-sheathed cable   (b) open wiring   (c) knob-and-tube wiring   (d) all of these

89. Short sections of metal enclosures or raceways used to provide support or protection of _____ from physical damage are not required to be grounded.

    (a) conduit   (b) 600V feeders   (c) cable assemblies   (d) none of these

90. Bonding must be provided where necessary to ensure _____ and the capacity to conduct safely any fault current likely to be imposed.

    (a) electrical continuity   (b) fiduciary responsibility   (c) listing requirements   (d) electrical demand

91. The noncurrent-carrying metal parts of service equipment, such as _____, must be effectively bonded together.

    (a) service raceways, cable trays, or service cable armor
    (b) service equipment enclosures containing service conductors, including meter fittings, boxes, or the like, interposed in the service raceway or armor
    (c) the metallic raceway or armor enclosing a grounding electrode conductor
    (d) all of these

92. Service equipment, service raceways, and service conductor enclosures must be bonded _____.

    (a) to the grounded service conductor
    (b) by threaded raceways into enclosures, couplings, hubs, conduit bodies, etc.
    (c) by listed bonding devices with bonding jumpers
    (d) any of these

93. Service raceways threaded into metal service equipment such as bosses (hubs) are considered to be effectively _____ to the service metal enclosure.

    (a) attached   (b) bonded   (c) grounded   (d) none of these

94. Service metal raceways and metal clad cables are considered effectively bonded when using threadless couplings and connectors that are _____.

    (a) nonmetallic   (b) made up tight
    (c) sealed   (d) these are never allowed for bonding

95. Bonding jumpers must be used around _____ knockouts that are punched or otherwise formed so as to impair the electrical connection to ground. Standard locknuts or bushings cannot be the sole means for this bonding.

    (a) concentric   (b) eccentric   (c) field-punched   (d) a or b

96. An accessible means external to enclosures for connecting intersystem _____ conductors must be provided at the service equipment and at the disconnecting means.

    (a) bonding   (b) grounding   (c) secondary   (d) a and b

(• Indicates that 75% or fewer exam takers get the question correct)

97. •Metal raceways, cable trays, cable armor, cable sheath, enclosures, frames, fittings, and other metal noncurrent-carrying parts that serve as the grounding conductor must be _____ where necessary to ensure electrical continuity and to have the capacity to conduct safely any fault current likely to be imposed.

    (a) grounded   (b) effectively bonded   (c) soldered or welded   (d) any of these

98. When bonding enclosures, metal raceways, frames, fittings, and other metal noncurrent-carrying parts, any nonconductive paint, enamel, or similar coating must be removed at _____.

    (a) contact surfaces   (b) threads   (c) contact points   (d) all of these

99. Where required to reduce electric noise for electronic equipment, electrical continuity of the metal raceway is not required and the metal raceway can terminate to a(n) _____ nonmetallic fitting(s) or spacer on the electronic equipment.

    (a) listed   (b) labeled   (c) identified   (d) marked

100. For circuits over 250 volts-to-ground (277/480V), electrical continuity can be maintained between a box or enclosure where no oversized, concentric, or eccentric knockouts are encountered, and a metal conduit by _____.

    (a) threadless fittings for cables with metal sheath
    (b) double locknuts on threaded conduit (one inside and one outside the box or enclosure)
    (c) fittings that have shoulders that seat firmly against the box with a locknut on the inside or listed fittings identified for the purpose.
    (d) all of these

101. When metal raceways and cables with metal sheaths are connected to enclosures at oversized, concentric, or eccentric knockouts for circuits over 250 volts to ground that do not contain service conductors, the electrical continuity of the raceway or metal cable sheath must be ensured by bonding similar to the requirements for service raceways.

    (a) True   (b) False

102. Regardless of the voltage of the electrical system, the electrical continuity of noncurrent-carrying metal parts of equipment, raceways, and other enclosures in any hazardous (classified) location as defined in Article 500 must be ensured by any of the methods specified in 250.92(B)(2) through (B)(4). One or more of these _____ methods must be used whether or not supplementary equipment grounding conductors are installed.

    (a) grounded   (b) securing   (c) sealing   (d) bonding

103. Equipment bonding jumpers must be of copper or other corrosion-resistant material. A bonding jumper must be a _____ or similar suitable conductor.

    (a) wire   (b) bus   (c) screw   (d) any of these

104. Equipment bonding jumpers on the supply side of the service must be no smaller than the sizes shown in _____.

    (a) Table 250.66   (b) Table 250.122   (c) Table 310.16   (d) Table 310.15(B)(6)

105. Where service entrance conductors are paralleled in two or more raceways or cables, the bonding jumper for each raceway or cable must be based on the size of the _____ in each raceway or cable.

    (a) overcurrent protection for conductors
    (b) grounded neutral conductors
    (c) service entrance conductors
    (d) sum of all conductors in the raceway

(• Indicates that 75% or fewer exam takers get the question correct)

106. A service is supplied by three metal raceways. Each raceway contains 600 kcmil ungrounded (phase) conductors. Determine the size of the service bonding jumper for each raceway.

(a) 1/0 AWG (b) 2/0 AWG (c) 225 kcmil (d) 500 kcmil

107. The equipment bonding jumper on the supply side of services (service raceway) must be sized according to the _____.

(a) calculated load
(b) service-entrance conductor size
(c) service-drop size
(d) load to be served

108. What is the minimum size copper bonding jumper for a service raceway containing 4/0 THHN aluminum conductors?

(a) 6 AWG aluminum (b) 3 AWG copper (c) 4 AWG aluminum (d) 4 AWG copper

109. What is the minimum size copper equipment bonding jumper required for equipment connected to a 40A circuit?

(a) 12 AWG (b) 14 AWG (c) 8 AWG (d) 10 AWG

110. The equipment bonding jumper can be installed on the outside of a raceway providing the length of the run is not more than _____ and the bonding jumper is routed with the raceway.

(a) 12 in. (b) 24 in. (c) 36 in. (d) 72 in.

111. The metal water-piping system(s) must be bonded to the _____.

(a) grounded neutral conductor at the service
(b) service equipment enclosure
(c) equipment grounding bar or bus at any panelboard within the building
(d) a or b

112. The bonding jumper used to bond the metal water piping system to the service must be sized in accordance with _____.

(a) Table 250.66 (b) Table 250.122 (c) Table 310.16 (d) Table 310.15(B)(6)

113. A building or structure that is supplied by a feeder must have the interior metal water-piping system bonded with a conductor sized from _____.

(a) Table 250.66 (b) Table 250.122 (c) Table 310.16 (d) none of these

114. Metal gas piping is considered bonded by the circuit's equipment grounding conductor of the circuit that is likely to energize the piping.

(a) True (b) False

115. Exposed structural metal that is interconnected to form a steel building frame, that is not intentionally grounded and is likely to become energized, must be bonded to:

(a) The service equipment enclosure.
(b) The grounded neutral conductor at the service.
(c) The grounding electrode conductor where of sufficient size.
(d) any of these

116. The metal water-pipe system of a building or structure is not required to be bonded to the separately derived system neutral terminal where the metal frame of the building or structure is used as the grounding electrode for the separately derived system and is bonded to the metal water piping in the area served by the separately derived system.

(a) True (b) False

(• Indicates that 75% or fewer exam takers get the question correct)

117. Lightning protection system ground terminals _____ be bonded to the building grounding electrode system.

(a) must (b) must not (c) can (d) none of these

118. Metal raceways, enclosures, frames, and other noncurrent-carrying metal parts of electric equipment installed on a building equipped with a lightning protection system may require spacing from the lightning protection conductors, typically 6 ft through air or ___ through dense materials, such as concrete, brick, wood, etc.

(a) 2 ft (b) 3 ft (c) 4 ft (d) 6 ft

119. Exposed noncurrent-carrying metal parts of fixed equipment likely to become energized must be grounded where _____.

(a) within 8 ft vertically or 5 ft horizontally of ground or grounded metal objects
(b) located in wet or damp locations and not isolated
(c) in electrical contact with metal
(d) any of these

120. Which of the following appliances installed in residential occupancies need not be grounded?

(a) Toaster (b) Aquarium (c) Dishwasher (d) Refrigerator

121. An equipment grounding conductor run with, or enclosing, the circuit conductors must be _____ or metal raceway as listed in 250.118.

(a) a copper conductor (b) an aluminum conductor (c) a copper-clad aluminum conductor (d) any of these

122. Flexible metal conduit can be used as the equipment grounding conductor if the length in any ground return path does not exceed 6 ft and the circuit conductors contained in the conduit are protected by overcurrent devices rated at _____ or less.

(a) 15A (b) 20A (c) 30A (d) 60A

123. For flexible metal conduit (FMC) and liquidtight flexible metal conduit (LFMC), an equipment grounding conductor is required regardless of the size of the overcurrent protection if the FMC or LFMC is installed for the reason of _____.

(a) physical protection (b) flexibility (c) protection from moisture (d) communications systems

124. Liquidtight flexible metal conduit (LFMC) up to 1/2 in. trade size can be used as the equipment grounding conductor if the length in any ground return path does not exceed 6 ft and the circuit conductors contained in the conduit are protected by overcurrent devices rated at _____ or less when the conduit is not installed for flexibility.

(a) 15A (b) 20A (c) 30A (d) 60A

125. Liquidtight flexible metal conduit (LFMC) in ¾ through 1¼ in. trade sizes can be used as the equipment grounding conductor if the length in any ground return path does not exceed 6 ft and the circuit conductors contained in the conduit are protected by overcurrent devices rated at _____ or less when the conduit is not installed for flexibility.

(a) 15A (b) 20A (c) 30A (d) 60A

126. An equipment grounding conductor must be identified by _____.

(a) a continuous outer finish that is green
(b) being bare
(c) a continuous outer finish that is green with one or more yellow stripes
(d) any of these

(• Indicates that 75% or fewer exam takers get the question correct)

127. Conductors with insulation that is _____ cannot be used for ungrounded or grounded neutral conductors.

    (a) green
    (b) green with one or more yellow stripes
    (c) a or b
    (d) white

128. Where an equipment grounding conductor consists of a raceway, cable tray, cable armor, cablebus framework, or cable sheath, it must be installed _____.

    (a) in accordance with applicable *Code* provisions
    (b) using fittings for joints and terminations approved for the use
    (c) with all connections, joints, and fittings made tight using suitable tools
    (d) all of these

129. Equipment grounding conductors for feeder taps must be sized in accordance with _____ based on the ampere rating of the circuit protection device ahead of the feeder, but in no case is it required to be larger than the circuit conductors.

    (a) Table 250.66   (b) Table 250.94   (c) Table 250.122   (d) Table 220.19

130. Equipment grounding conductors must be the same size as the circuit conductors for _____ circuits.

    (a) 15A   (b) 20A   (c) 30A   (d) all of these

131. The equipment grounding conductor must not be smaller than shown in Table 250.122, but it must not be required to be larger than the circuit conductors supplying the equipment.

    (a) True   (b) False

132. When ungrounded conductors are increased in size, the equipment grounding conductor is not required to be increased because it is not a current-carrying conductor.

    (a) True   (b) False

133. •What size equipment grounding conductor is required for a nonmetallic raceway that contains the following three circuits?

    Circuit 1 - 12 AWG protected by a 20A device
    Circuit 2 - 10 AWG protected by a 30A device
    Circuit 3 - 8 AWG protected by a 40A device

    (a) 10 AWG   (b) 6 AWG   (c) 8 AWG   (d) 12 AWG

134. When a single equipment grounding conductor is used for multiple circuits in the same raceway or cable, the single equipment grounding conductor must be sized according to _____.

    (a) the combined rating of all the overcurrent protection devices
    (b) the largest overcurrent protection device of the multiple circuits
    (c) the combined rating of all the loads
    (d) any of these

135. Where conductors are run in parallel in multiple raceways or cables, the equipment grounding conductor, where used, must be run in parallel in each raceway or cable.

    (a) True   (b) False

(• Indicates that 75% or fewer exam takers get the question correct)

136. The terminal of a wiring device for the connection of the equipment grounding conductor must be identified by a green-colored, _____.

   (a) not readily removable terminal screw with a hexagonal head
   (b) hexagonal, not readily removable terminal nut
   (c) pressure wire connector
   (d) any of these

137. A grounding-type receptacle can replace a nongrounding-type receptacle at an outlet box that does not contain an equipment grounding conductor if the equipment grounding conductor is connected to the _____.

   (a) grounding electrode system as described in 250.50
   (b) grounding electrode conductor
   (c) equipment grounding terminal bar within the enclosure where the branch circuit for the receptacle originates
   (d) any of these

138. Cord-and-Plug connected equipment must be grounded by means of _____.

   (a) an equipment grounding conductor in the cable assembly
   (b) a separate flexible wire or strap
   (c) either a or b
   (d) none of these

139. Ranges and clothes dryers for existing branch circuit installations that were installed with the frame grounded by the grounded circuit conductor are allowed to continue this practice if all conditions of the exception to 250.140 are met.

   (a) True   (b) False

140. The grounded circuit conductor is permitted to ground noncurrent-carrying metal parts of equipment, raceways, and other enclosures at the supply side or within the enclosure of the ac service-disconnecting means.

   (a) True   (b) False

141. A grounded circuit conductor must not be used for grounding noncurrent-carrying metal parts of equipment on the load side of _____.

   (a) the service disconnecting means
   (b) the separately derived system disconnecting means
   (c) overcurrent protection devices for separately derived systems not having a main disconnecting means
   (d) all of these

142. On the load side of the service disconnecting means, the _____ circuit conductor is permitted to ground meter enclosures if all meter enclosures are located near the service disconnecting means and no service ground-fault protection is installed.

   (a) grounding   (b) bonding   (c) grounded   (d) phase

143. An _____ must be used to connect the grounding terminal of a grounding-type receptacle to a grounded box.

   (a) equipment bonding jumper   (b) equipment grounding jumper
   (c) a or b   (d) a and b

(• Indicates that 75% or fewer exam takers get the question correct)

144. Where a metal box is surface-mounted, direct metal-to-metal contact between the device yoke and the box is permitted to ground the receptacle to the box. Unless the receptacle is listed as _____, at least one of the insulating retaining washers must be removed from the receptacle to ensure direct metal-to-metal contact between the device yoke and metal outlet box.

    (a) self-grounding   (b) weatherproof   (c) metal contact sufficient   (d) isolated grounding

145. An equipment bonding jumper must be used to connect the grounding terminal of a grounding-type receptacle to a grounded box. Where the box is surface-mounted, direct metal-to-metal contact between the device yoke and the box can be permitted to ground the receptacle to the box.

    (a) True   (b) False

146. For a cover mounted receptacle, direct metal-to-metal contact of the receptacle yoke and the metal cover is always considered to be sufficiently bonded and no equipment bonding jumper is required.

    (a) True   (b) False

147. Receptacle yokes designed and _____ as self-grounding are permitted to establish the bonding path between the device yoke and a grounded outlet box.

    (a) approved   (b) advertised   (c) listed   (d) installed

148. An equipment bonding jumper for a grounding-type receptacle must be installed between the receptacle and a flush-mounted outlet box, even when the contact device is listed as self-grounding.

    (a) True   (b) False

149. Contact devices or yokes designed and listed as self-grounding are permitted in conjunction with the supporting screws to establish the grounding circuit between the device yoke and flush-type boxes.

    (a) True   (b) False

150. Equipment bonding jumpers are not required for receptacles listed as self-grounding that have mounting screws to provide the grounding continuity between the metal yoke and the flush box.

    (a) True   (b) False

151. Where circuit conductors are spliced within a box, or terminated on equipment within or supported by a box, any equipment grounding conductors associated with those circuit conductors must be spliced or joined within the box or to the box with devices suitable for the use.

    (a) True   (b) False

152. •Where circuit conductors are spliced within a box, or terminated on equipment within or supported by a box, any equipment grounding conductors associated with those circuit conductors must be spliced or joined in the box or to the box with devices suitable for the use. This does not apply to insulated equipment grounding conductors for isolated ground receptacles for electronic equipment.

    (a) True   (b) False

153. When equipment grounding conductor(s) are installed in a metal box, an electrical connection is required between the equipment grounding conductor and the metal box enclosure by means of a _____.

    (a) grounding screw   (b) soldered connection   (c) listed grounding device   (d) a or c

(• Indicates that 75% or fewer exam takers get the question correct)

Article 280 Surge Arresters

Article Overview

This article covers general requirements, installation requirements, and connection requirements for surge arresters installed on premises wiring systems. Surge arresters are generally installed on the supply side (line side) of the service disconnecting means [280.22]. Some surge arresting devices are dual-listed for use on either the line side or load side of the service disconnecting means.

Transient voltage surges are short-term deviations from a desired voltage level, of high enough magnitude to cause equipment malfunction or damage. Surge arresters are designed to reduce transient voltages present on utility power lines and other line-side equipment. A surge arrester limits surge voltages by discharging or bypassing surge current. It prevents continued flow of follow current without sacrificing itself in the process so it can repeatedly provide the intended protection.

Questions

1. A surge arrester is a protective device for limiting surge voltages by _____ or bypassing surge current.

    (a) decreasing (b) discharging (c) limiting (d) derating

2. Line and ground-connecting conductors for a surge arrester must not be smaller than _____ AWG copper.

    (a) 14 (b) 12 (c) 10 (d) 8

3. The conductor between a surge arrester and the line and the grounding connection must not be smaller than _____ AWG copper for installations operating at 1 kV or more.

    (a) 4 (b) 6 (c) 8 (d) 2

(• Indicates that 75% or fewer exam takers get the question correct)

Article 285 Transient Voltage Surge Suppressors (TVSSs)

Article Overview

This article covers general requirements, installation requirements, and connection requirements for transient voltage surge suppressors (TVSSs) permanently installed on premises wiring systems. It doesn't apply to cord-and-plug connected units such as "computer power strips."

TVSS devices generally apply to the load side of the service disconnecting means [285.21]. Transient voltage surges are short-term deviations from a desired voltage level, of high enough magnitude to cause equipment malfunction or damage. TVSS devices are designed to reduce transient voltages present on premises power distribution wiring and load-side equipment, particularly sensitive electronic equipment such as computers, telecommunications equipment, security systems, and electronic appliances.

A TVSS is designed to *shunt* the current and *clamp* the voltage. This is a key difference between a TVSS and a surge arrester. A TVSS clamps at a given voltage, depending on its rating, regardless of the actual surge current at its input terminals. This allows a TVSS to provide a more precise level of protection than a surge arrester can.

Questions

1. Article 285 covers surge arresters.

   (a) True (b) False

2. The scope of Article 285 applies to devices such as cord-and-plug connected TVSS units or receptacles, or appliances that have integral TVSS protection.

   (a) True (b) False

3. A TVSS is listed to limit transient voltages by diverting or limiting surge current.

   (a) True (b) False

4. A TVSS device must be listed.

   (a) True (b) False

5. TVSSs must be marked with their short-circuit current rating, and they must not be installed where the available fault current is in excess of that rating.

   (a) True (b) False

6. The conductors for the TVSS cannot be any longer than _____, and unnecessary bends should be avoided.

   (a) 6 in. (b) 12 in. (c) 18 in. (d) none of these

7. A TVSS can be connected anywhere on the premises wiring system.

   (a) True (b) False

(• Indicates that 75% or fewer exam takers get the question correct)

Questions for Chapter 3—Wiring Methods and Materials

Article 300 Wiring Methods

Article Overview

Article 300 contains the general requirements for all wiring methods included in the *NEC*. However, this article doesn't apply to signaling and communications systems, which are covered in Chapters 7 and 8, except as specifically referenced in those chapters.

This article is primarily concerned with how you install, route, splice, protect, and secure conductors and raceways. How well you conform to the requirements of Article 300 will generally be evident in the finished work, because many of the requirements tend to determine the appearance of the installation.

Because of this, it's often easy to spot Article 300 problems if you're looking for *Code* violations. For example, you can easily spot when someone runs an equipment grounding (bonding) conductor outside a raceway instead of grouping all conductors of a circuit together, as required by 300.3(B). From your study of Article 300 you'll learn there's a difference between running an equipment grounding (bonding) conductor outside a raceway, which is illegal, and running an external bonding jumper outside the same raceway.

This is just one of the common points of confusion your studies here will clear up for you. To help achieve that end, be sure to carefully consider the illustrations included in the accompanying textbook and also refer to Article 100 as needed.

Questions

1. Single conductors as specified in Table 310.13 are only permitted when installed as part of a recognized wiring method from Chapter _____ of the *NEC*.

    (a) 4   (b) 3   (c) 2   (d) 9

2. All conductors of a circuit, including the grounded and equipment grounding conductors, must be contained within the same _____.

    (a) raceway   (b) cable   (c) trench   (d) all of these

3. •Circuit conductors that operate at 277V (with 600V insulation) may occupy the same enclosure or raceway with 48V dc conductors that have an insulation rating of 300V.

    (a) True   (b) False

(• Indicates that 75% or fewer exam takers get the question correct)

4. In both exposed and concealed locations, where a cable or nonmetallic raceway-type wiring method is installed through bored holes in joists, rafters, or wood members, holes must be bored so that the edge of the hole is _____ the nearest edge of the wood member.

   (a) not less than 1¼ inch from
   (b) immediately adjacent to
   (c) not less than 1/16 inch from
   (d) 90° away from

5. When unable to maintain the minimum required distance from the edge of a wood framing member to a bored hole for cable or nonmetallic raceway installation, the cable or raceway must be protected from penetration by screws or nails by a steel plate or bushing at least _____ and of appropriate length and width to cover the area of the wiring. A thinner plate that provides equal or better protection may be used if listed and marked.

   (a) ¼ in. thick
   (b) ⅛ in. thick
   (c) 1/16 in. thick
   (d) 24 gauge

6. Cables laid in wood notches require protection against nails or screws by using a steel plate at least _____ thick, installed before the building finish is applied. A thinner plate that provides equal or better protection may be used if listed and marked.

   (a) 1/16 in.
   (b) ⅛ in.
   (c) ½ in.
   (d) none of these

7. Where NM cables pass through cut or drilled slots or holes in metal members, the cable needs to be protected by _____ securely covering all metal edges fastened in the opening prior to installation of the cable.

   (a) listed bushings
   (b) listed grommets
   (c) plates
   (d) a or b

8. Where NM cable passes through factory or field openings in metal members, it must be protected by _____ bushings or _____ grommets that cover metal edges. The protection fitting must be securely fastened in the opening prior to the installation of the cable.

   (a) approved
   (b) identified
   (c) listed
   (d) none of these

9. Where nails or screws are likely to penetrate nonmetallic-sheathed cable or electrical nonmetallic tubing installed through metal framing members, a steel sleeve, steel plate, or steel clip not less than _____ in thickness must be used to protect the cable or tubing. A thinner plate that provides equal or better protection may be used if listed and marked.

   (a) 1/16 in.
   (b) ⅛ in.
   (c) ½ in.
   (d) none of these

10. Wiring methods installed behind panels that allow access, such as the space above a dropped ceiling, are required to be _____ according to their applicable Articles.

    (a) supported
    (b) painted
    (c) in a metal raceway
    (d) all of these

11. In both exposed and concealed locations, where a cable or nonmetallic raceway-type wiring method is installed parallel to framing members such as joists, rafters, or studs or furring strips, the nearest outside surface of the cable or raceway must be _____ the nearest edge of the framing member where nails or screws are likely to penetrate.

    (a) not less than 1¼ inch from
    (b) immediately adjacent to
    (c) not less than 1/16 inch from
    (d) 90° away from

12. When unable to maintain the minimum required distance from the edge of a wood framing member when installing a cable or nonmetallic raceway parallel to framing members, the cable or raceway must be protected from penetration by screws or nails by a steel plate or bushing at least _____ and of appropriate length and width to cover the area of the wiring. A thinner plate that provides equal or better protection may be used if listed and marked.

    (a) ¼ in. thick
    (b) ⅛ in. thick
    (c) 1/16 in. thick
    (d) 24 gauge

(• Indicates that 75% or fewer exam takers get the question correct)

13. Cable or nonmetallic raceway-type wiring methods installed in a groove, to be covered by wallboard, siding, paneling, carpeting, or similar finish must be protected by 1/16 in. thick _____ or by not less than 1¼ in. of free space for the full length of the groove. A thinner plate that provides equal or better protection may be used if listed and marked.

    (a) steel plate (b) steel sleeve (c) PVC bushing (d) a or b

14. Where underground conductors and cables emerge from underground, they must be protected by enclosures or raceways to a point _____ above finished grade. In no case can the protection be required to exceed 18 in. below grade.

    (a) 3 ft (b) 6 ft (c) 8 ft (d) 10 ft

15. The _____ is defined as the area between the top of direct-burial cable and the finished grade.

    (a) notch (b) cover (c) gap (d) none of these

16. What is the minimum cover requirement in inches for direct burial UF cable installed outdoors that supplies power to a 120V, 30A circuit?

    (a) 6 in. (b) 12 in. (c) 18 in. (d) 24 in.

17. Rigid metal conduit that is directly buried outdoors must have at least _____ of cover.

    (a) 6 in. (b) 12 in. (c) 18 in. (d) 24 in.

18. •When installing raceways underground in rigid nonmetallic conduit and other approved raceways, there must be a minimum of _____ of cover.

    (a) 6 in. (b) 12 in. (c) 18 in. (d) 22 in.

19. •What is the minimum cover requirement in inches for UF cable supplying power to a 120V, 15A GFCI-protected circuit outdoors under a driveway of a one-family dwelling?

    (a) 12 in. (b) 24 in. (c) 16 in. (d) 6 in.

20. UF cable used with a 24V landscape lighting system is permitted to have a minimum cover of _____.

    (a) 6 in. (b) 12 in. (c) 18 in. (d) 24 in.

21. Service conductors that are not encased in concrete and that are buried 18 in. or more below grade must have their location identified by a warning ribbon placed in the trench at least _____ above the underground installation.

    (a) 6 in. (b) 12 in. (c) 18 in. (d) none of these

22. Direct-buried conductors or cables can be spliced or tapped without the use of splice boxes when the splice or tap is made in accordance with 110.14(B).

    (a) True (b) False

23. Backfill used for underground wiring must not _____.

    (a) damage the wiring method (b) prevent compaction of the fill
    (c) contribute to the corrosion of the raceway (d) all of these

24. Conduits or raceways through which moisture may contact live parts must be _____ at either or both ends.

    (a) sealed (b) plugged (c) bushed (d) a or b

(• Indicates that 75% or fewer exam takers get the question correct)

25. •When installing direct-buried cables, a _____ must be used at the end of a conduit that terminates underground.

    (a) splice kit (b) terminal fitting (c) bushing (d) b or c

26. All conductors of the same circuit are required to be _____.

    (a) in the same raceway or cable
    (b) in close proximity in the same trench
    (c) the same size
    (d) a or b

27. Direct buried conductors, cables, or raceways, which are subject to movement by settlement or frost, must be arranged to prevent damage to the _____ or to equipment connected to the raceways.

    (a) siding of the building mounted on
    (b) landscaping around the cable or raceway
    (c) the enclosed conductors
    (d) expansion fitting

28. Cables or raceways installed using directional boring equipment must be _____ for this purpose.

    (a) marked (b) listed (c) labeled (d) approved

29. Raceways, cable trays, cable bus, auxiliary gutters, cable armor, boxes, cable sheathing, cabinets, elbows, couplings, fittings, supports, and support hardware must be of materials suitable for _____.

    (a) corrosive locations
    (b) wet locations
    (c) the environment in which they are to be installed
    (d) none of these

30. Which of the following metal parts must be protected from corrosion both inside and out?

    (a) Ferrous metal raceways (b) Metal elbows (c) Boxes (d) all of these

31. Where corrosion protection is necessary and the conduit is threaded in the field, the threads must be coated with a(n) _____, electrically conductive, corrosion-resistance compound.

    (a) marked (b) listed (c) labeled (d) approved

32. Metal raceways, boxes, fittings, supports, and support hardware can be installed in concrete or in direct contact with the earth or other areas subject to severe corrosive influences, where _____ approved for the conditions, or where provided with corrosion protection approved for the purpose.

    (a) the soil is (b) made of material (c) the qualified installer is (d) none of these

33. Nonferrous raceways, cable trays, cablebus, auxiliary gutters, cable armor, boxes, cable sheathing, cabinets, elbows, couplings, nipples, fittings, supports, and support hardware _____ must be provided with supplementary corrosion protection.

    (a) embedded or encased in concrete
    (b) in direct contact with the earth
    (c) likely to become energized
    (d) a or b

34. Nonmetallic raceways, cable trays, cablebus, auxiliary gutters, boxes, cables with a nonmetallic outer jacket and internal metal armor or jacket, cable sheathing, cabinets, elbows, couplings, nipples, fittings, supports and support hardware must be made of material _____.

    (a) listed for the condition
    (b) approved for the condition
    (c) both a and b
    (d) either a or b

(• Indicates that 75% or fewer exam takers get the question correct)

35. Nonmetallic raceways, cable trays, cablebus, auxiliary gutters, boxes, and cables with nonmetallic outer jackets must be made of material approved for the conditions where they will be installed, and where exposed to sunlight, the materials must be _____.

   (a) listed as sunlight resistant
   (b) identified as sunlight resistant
   (c) both a and b
   (d) either a or b

36. Nonmetallic raceways, cable trays, cablebus, auxiliary gutters, boxes, and cables with nonmetallic outer jackets must be made of material approved for the conditions where they will be installed and where exposed to chemicals, the materials or coatings must be _____.

   (a) listed as inherently resistant to chemicals
   (b) identified for the specific chemical reagent
   (c) both a and b
   (d) either a or b

37. An exposed wiring system for indoor wet locations where walls are frequently washed must be mounted so that there is at least _____ between the mounting surface and the electrical equipment.

   (a) a ¼ in. airspace
   (b) separation by insulated bushings
   (c) separation by noncombustible tubing
   (d) none of these

38. Where portions of a cable raceway or sleeve are subjected to different temperatures and where condensation is known to be a problem, as in cold storage areas of buildings or where passing from the interior to the exterior of a building, the _____ must be filled with an approved material to prevent the circulation of warm air to a colder section of the raceway or sleeve.

   (a) raceways (b) sleeve (c) a or b (d) none of these

39. Raceways must be provided with expansion fittings where necessary to compensate for thermal expansion and contraction.

   (a) True (b) False

40. Raceways or cable trays containing electric conductors must not contain any pipe, tube, or equal for steam, water, air, gas, drainage, or any service other than _____.

   (a) allowed by the authority having jurisdiction
   (b) electrical
   (c) pneumatic
   (d) designed by the engineer

41. •Metal raceways, cable armor, and other metal enclosures for conductors must be _____ joined together to form a continuous electrical conductor.

   (a) electrically (b) permanently (c) metallically (d) none of these

42. Raceways, cable assemblies, boxes, cabinets, and fittings must be securely fastened in place. Support wires and associated fittings that provide secure support and that are installed in addition to the ceiling grid support wires are permitted as the sole support.

   (a) True (b) False

43. Where independent support wires of a ceiling assembly are used to support raceways, cable assemblies, or boxes above a ceiling, they must be secured at both ends. Cables and raceways must _____.

   (a) be identified for this purpose
   (b) not be supported by ceiling grids
   (c) not contain conductors larger than 14 AWG
   (d) be identified by painting them orange

44. Electrical wiring within the cavity of a fire-rated floor-ceiling or roof-ceiling assembly cannot be supported by the ceiling assembly or ceiling support wires. An independent means of support must be provided which _____.

   (a) is permitted to be attached to the ceiling assembly
   (b) cannot be attached to the ceiling assembly
   (c) can be nonmetallic material
   (d) none of these

(• Indicates that 75% or fewer exam takers get the question correct)

45. The independent support wires for wiring in a fire-rated ceiling assembly must be distinguishable from fire-rated suspended-ceiling framing support wires by _____.

    (a) color (b) tagging (c) other effective means (d) any of these

46. Ceiling-support wires used for the support of electrical raceways and cables within nonfire rated assemblies are required to be distinguishable from the suspended-ceiling framing support wires.

    (a) True (b) False

47. Raceways are allowed to be used as a means of support when the raceway contains electrical power supply conductors for electrically controlled equipment and the raceway is used to support Class 2 circuit conductors or cables that connect to the same equipment.

    (a) True (b) False

48. Cable wiring methods must not be used as a means of support for _____.

    (a) other cables (b) raceways (c) non electrical equipment (d) a, b, or c

49. Metal or nonmetallic raceways, cable armors, and cable sheaths _____ between cabinets, boxes, fittings or other enclosures or outlets.

    (a) can be attached with electrical tape (b) are allowed gaps for expansion
    (c) must be continuous (d) none of these

50. Conductors in raceways must be _____ between outlets, boxes, devices, and so forth.

    (a) continuous (b) installed (c) copper (d) in conduit

51. In multiwire circuits, the continuity of the _____ conductor must not be dependent upon the device connections.

    (a) ungrounded (b) grounded (c) grounding (d) a and b

52. An 8 x 8 x 4 in. deep junction/splice box requires 6 in. of free conductor, measured from the point in the box where the conductors enter the enclosure. The 3 in. of conductor outside-the-box rule _____.

    (a) does apply (b) does not apply (c) sometimes applies (d) none of these

53. When the opening to an outlet, junction, or switch point is less than 8 in. in any dimension, each conductor must be long enough to extend at least _____ outside the opening of the enclosure.

    (a) 0 in. (b) 3 in. (c) 6 in. (d) 12 in.

54. Fittings and connectors must be used only with the specific wiring methods for which they are designed and listed.

    (a) True (b) False

55. A box or conduit body is not required where cables enter or exit from conduit or tubing that is used to provide cable support or protection against physical damage. A fitting must be provided on the end(s) of the conduit or tubing to _____.

    (a) allow for the future connection of a box (b) be used for a future pull point
    (c) protect the cable from abrasion (d) allow the coupling of another section of conduit

(• Indicates that 75% or fewer exam takers get the question correct)

56. A box or conduit body is not required for splices and taps in direct-buried conductors and cables as long as the splice is made with a splicing device that is identified for the purpose.

    (a) True (b) False

57. Splices and taps are permitted in cabinets or cutout boxes if the conductors, splices, and taps do not fill the wiring space at any cross-section to more than _____ percent.

    (a) 20 (b) 40 (c) 60 (d) 75

58. Where accessible only to qualified persons, a box or conduit body is not required for conductors in _____ when installed in accordance with applicable *Code* provisions.

    (a) manholes (b) handhole enclosures (c) a or b (d) elevator pits

59. A bushing is permitted in lieu of a box or terminal where conductors emerge from a raceway and enter or terminate at equipment, such as open switchboards, unenclosed control equipment, or similar equipment.

    (a) True (b) False

60. A _____ is permitted in lieu of a box or terminal fitting at the end of a conduit where the raceway terminates behind an unenclosed switchboard or similar equipment.

    (a) bushing (b) bonding bushing (c) coupling (d) connector

61. The number of conductors permitted in a raceway must be limited to _____.

    (a) permit heat to dissipate
    (b) prevent damage to insulation during installation
    (c) prevent damage to insulation during removal of conductors
    (d) all of these

62. Raceways must be _____ between outlet, junction, or splicing points prior to the installation of conductors.

    (a) installed complete
    (b) tested for ground faults
    (c) a minimum of 80 percent completed
    (d) none of these

63. Prewired raceway assemblies are permitted only where specifically permitted in the *Code* for the applicable wiring method.

    (a) True (b) False

64. Short sections of raceways used for _____ are not required to be installed complete between outlet, junction, or splicing points.

    (a) meter to service enclosure connection
    (b) protection of cables from physical damage
    (c) nipples
    (d) separately derived systems

65. Metal raceways must not be _____ by welding to the raceway unless specifically designed to be, or otherwise specifically permitted to be, by the *Code*.

    (a) supported (b) terminated (c) connected (d) all of these

66. A 100 ft vertical run of 4/0 AWG copper requires the conductors to be supported at _____ locations.

    (a) 4 (b) 5 (c) 2 (d) none of these

67. A vertical run of 4/0 AWG copper must be supported at intervals not exceeding _____.

    (a) 80 ft (b) 100 ft (c) 120 ft (d) 40 ft

(• Indicates that 75% or fewer exam takers get the question correct)

68. Conductors in metal raceways and enclosures must be so arranged as to avoid heating the surrounding metal by alternating-current induction. To accomplish this, the _____ conductor(s) must be grouped together.

    (a) phase   (b) grounded   (c) ungrounded   (d) all of these

69. _____ is a nonferrous, nonmagnetic metal that has no heating due to inductive hysteresis heating.

    (a) Steel   (b) Iron   (c) Aluminum   (d) all of these

70. Electrical installations in hollow spaces, vertical shafts, and ventilation or air-handling ducts must be made so that the possible spread of fire or products of combustion will not be _____.

    (a) substantially increased   (b) allowed   (c) inherent   (d) possible

71. Openings around electrical penetrations through fire-resistant-rated walls, partitions, floors, or ceilings must _____ to maintain the fire resistance rating.

    (a) be documented   (b) not be allowed
    (c) be firestopped using approved methods   (d) be enlarged

72. No wiring of any type can be installed in ducts used to transport _____.

    (a) dust   (b) flammable vapors   (c) loose stock   (d) all of these

73. Equipment and devices are permitted within ducts or plenum chambers used to transport environmental air only if necessary for their direct action upon, or sensing of, the _____.

    (a) contained air   (b) air quality   (c) air temperature   (d) none of these

74. Type AC cable can be installed in ducts or plenums that are used for environmental air.

    (a) True   (b) False

75. One wiring method that is permitted in ducts or plenums used for environmental air is _____.

    (a) flexible metal conduit of any length   (b) electrical metallic tubing
    (c) armored cable (Type AC)   (d) nonmetallic-sheathed cable

76. When equipment or devices are installed in ducts or plenum chambers used to transport environmental air, and illumination is necessary to facilitate maintenance and repair, enclosed _____-type luminaires are permitted.

    (a) screw   (b) plug   (c) gasketed   (d) neon

77. •Wiring methods permitted in the hung ceiling area used for environmental air include _____.

    (a) electrical metallic tubing   (b) flexible metal conduit of any length
    (c) rigid metal conduit without an overall nonmetallic covering   (d) all of these

78. Electric wiring in the air-handling area beneath raised floors for data-processing systems is permitted in accordance with Article 645.

    (a) True   (b) False

79. Wiring methods and equipment installed behind panels designed to permit access (such as suspended ceiling panels) must be so arranged and secured so as to allow the removal of panels and access to the electrical equipment.

    (a) True   (b) False

(• Indicates that 75% or fewer exam takers get the question correct)

Article 310 Conductors for General Wiring

Article Overview

This article contains the general requirements for conductors, such as insulation markings, ampacity ratings, and conditions of use. Article 310 doesn't apply to conductors that are part of cable assemblies, flexible cords, fixture wires, or to conductors that are an integral part of equipment [90.6, 300.1(B)].

People most often make errors in applying the ampacity tables contained in Article 310. If you study the explanations carefully, you'll avoid common errors such as applying Table 310.17 when you should be applying Table 310.16.

But why so many tables? Why does Table 310.17 list the ampacity of 6 THHN as 105A, yet Table 310.16 lists the same conductor as having an ampacity of only 75A? To answer that, go back to Article 100 and review the definition of ampacity. Notice the phrase "conditions of use." What these tables do is set a maximum current value at which you can ensure the installation won't undergo premature failure of the conductor insulation in normal use.

The designations THHN, THHW, RHH, and so on, describe insulation. Every type of insulation has a heat withstand limit. When current flows through a conductor it creates heat. How well the insulation around a conductor can dissipate that heat depends on factors such as whether that conductor is in free air or not. Think what happens to you if you put on a sweater, a jacket, and then a coat—all at the same time. You heat up. Your skin cannot dissipate heat with all this clothing on nearly as well as it dissipates heat in free air.

Conductors fail with age. That's why we conduct cable testing and take other measures to predict failure and replace certain conductors (for example, feeders or critical equipment conductors) while they are still within design specifications. But conductor failure takes decades under normal use and occurs slowly—and it's a maintenance issue. However, if a conductor is forced to exceed the ampacity listed in the appropriate table, failure happens much more rapidly—often catastrophically. Exceeding the allowable ampacity is a safety issue.

Questions

1. Conductors must be insulated except where specifically allowed by the *NEC* to be bare, such as for equipment grounding or bonding purposes.

    (a) True   (b) False

2. Where installed in raceways, conductors _____ AWG and larger must be stranded.

    (a) 10   (b) 6   (c) 8   (d) 4

3. In general, the minimum size phase, neutral, or grounded neutral conductor permitted for use in parallel installations is _____ AWG.

    (a) 10   (b) 1   (c) 1/0   (d) 4

4. Conductors smaller than 1/0 AWG can be connected in parallel to supply control power, provided _____.

    (a) they are all contained within the same raceway or cable
    (b) each parallel conductor has an ampacity sufficient to carry the entire load
    (c) the circuit overcurrent protection device rating does not exceed the ampacity of any individual parallel conductor
    (d) all of these

(• Indicates that 75% or fewer exam takers get the question correct)

5. When conductors are run in parallel, the currents should be evenly divided between the individual parallel conductors so that each conductor is evenly heated. This is accomplished by ensuring that each of the conductors within a parallel set has the same _____ and all conductors terminate in the same manner.

   (a) length   (b) material   (c) cross-sectional area   (d) all of these

6. Where _____ conductors are run in separate raceways or cables, the same number of conductors must be used in each raceway or cable.

   (a) parallel   (b) control   (c) communication   (d) aluminum

7. It is not the intent of 310.4 to require that conductors of one phase, neutral, or grounded circuit conductor be the same as those of another phase, neutral, or grounded circuit conductor to achieve _____.

   (a) polarity   (b) balance   (c) grounding   (d) none of these

8. The parallel conductors in each phase or grounded neutral conductor must _____.

   (a) be the same length and conductor material
   (b) have the same circular mil area and insulation type
   (c) be terminated in the same manner
   (d) all of these

9. The minimum size conductor permitted in any building for branch circuits under 600V is _____ AWG.

   (a) 14   (b) 12   (c) 10   (d) 8

10. •Insulated conductors used in wet locations must be _____.

    (a) moisture-impervious metal-sheathed
    (b) RHW, TW, THW, THHW, THWN, XHHW
    (c) listed for wet locations
    (d) any of these

11. Insulated conductors and cables exposed to the direct rays of the sun must be _____.

    (a) covered with insulating material that is listed or listed and marked sunlight resistant
    (b) listed and marked sunlight resistant
    (c) listed for sunlight resistance
    (d) any of these

12. Where conductors of different insulation are associated together, the limiting temperature of any conductor must not be exceeded.

    (a) True   (b) False

13. The _____ rating of a conductor is the maximum temperature, at any location along its length, that the conductor can withstand over a prolonged period of time without serious degradation.

    (a) ambient   (b) temperature   (c) maximum withstand   (d) short-circuit

14. There are four principal determinants of conductor operating temperature, one of which is _____ generated internally in the conductor as the result of load current flow.

    (a) friction   (b) magnetism   (c) heat   (d) none of these

15. Conductors installed in conduit exposed to direct sunlight in close proximity to rooftops have been shown, under certain conditions, to experience an increase in temperature of _____ °F above ambient temperature.

    (a) 70   (b) 10   (c) 30   (d) 40

(• Indicates that 75% or fewer exam takers get the question correct)

16. Which conductor has an insulation temperature rating of 90°C?

    (a) RH (b) RHW (c) THHN (d) TW

17. TFE-insulated conductors are manufactured in sizes from 14 through _____ AWG.

    (a) 2 (b) 1 (c) 2/0 (d) 4/0

18. Lettering on conductor insulation indicates its intended condition of use. THWN is rated _____.

    (a) 75°C (b) for wet locations (c) a and b (d) not enough information

19. THW insulation has a _____ rating when installed within electric-discharge lighting equipment, such as through fluorescent luminaires.

    (a) 60°C (b) 75°C (c) 90°C (d) none of these

20. The ampacities listed in the Tables of Article 310 are based on temperature alone and do not take _____ into consideration.

    (a) continuous loads (b) voltage drop (c) insulation (d) wet locations

21. The ampacity of a conductor can be different along the length of the conductor. The higher ampacity is permitted to be used beyond the point of transition for a distance no more than _____ ft, or no more than _____ percent of the circuit length figured at the higher ampacity, whichever is less.

    (a) 10, 20 (b) 20, 10 (c) 10, 10 (d) 15, 15

22. Where six current-carrying conductors are run in the same conduit or cable, the ampacity of each conductor must be adjusted to a factor of _____ percent of its value.

    (a) 90 (b) 60 (c) 40 (d) 80

23. Each current-carrying conductor of a paralleled set of conductors must be counted as a current-carrying conductor for the purpose of applying the adjustment factors of 310.15(B)(2)(a).

    (a) True (b) False

24. Conductor derating factors do not apply to conductors in nipples having a length not exceeding _____

    (a) 12 in. (b) 24 in. (c) 36 in. (d) 48 in.

25. The ampacity adjustment factors of Table 310.15(B)(2)(a) do not apply to AC or MC cable without an overall outer jacket, if which of the following conditions are met?

    (a) Each cable has not more than three current-carrying conductors.
    (b) The conductors are 12 AWG copper.
    (c) No more than 20 current-carrying conductors are bundled or stacked.
    (d) all of these

26. •When bare grounding conductors are allowed, their ampacities are limited to _____.

    (a) 60°C
    (b) 75°C
    (c) 90°C
    (d) those permitted for the insulated conductors of the same size

(• Indicates that 75% or fewer exam takers get the question correct)

27. In a balanced 120/208V, 4-wire, 3Ø system, the grounded neutral conductor will carry _____ amperes if the loads supplied are linear loads and no harmonic currents are present.

    (a) full load (b) zero (c) fault-current (d) none of these

28. Service and feeder conductors may be sized using Table 310.15(B)(6) for_____.

    (a) any kind of service under 400 amps
    (b) only multifamily dwelling services
    (c) only 120/240 volt, 3-wire, 1Ø services for individual dwelling units
    (d) commercial services only

29. For individual dwelling units of _____ dwellings, Table 310.15(B)(6) can be used to size 3-wire, 1Ø, 120/240V service or feeder conductors that serve as the main power feeder.

    (a) one-family (b) two-family (c) multifamily (d) any of these

30. In designing circuits, the current-carrying capacity of conductors should be corrected for heat at room temperatures above _____.

    (a) 30°F (b) 86°F (c) 94°F (d) 75°F

(• Indicates that 75% or fewer exam takers get the question correct)

Article 312 Cabinets, Cutout Boxes, and Meter Socket Enclosures

Article Overview

This article addresses the installation and construction specifications for the items mentioned in its title. In Article 310, we observed that you need different ampacities for the same conductor, depending on conditions of use. The same thing applies to these items—just in a different way. For example, you can't use just any enclosure in a wet area or in a hazardous (classified) location. The conditions of use impose special requirements for these situations.

For all such enclosures, certain requirements apply—regardless of the use. For example, you must cover any openings, protect conductors from abrasion, and allow sufficient bending room for conductors.

Part I is where you'll find the requirements most useful to the electrician in the field. Part II applies to manufacturers. If you use name brand components that are listed or labeled, you do not need to be concerned with Part II. However, if you are specifying custom enclosures, you need to be familiar with these requirements to ensure the authority having jurisdiction approves the enclosures.

Questions

1. Cabinets or cutout boxes installed in wet locations must be _____.

   (a) waterproof   (b) raintight   (c) weatherproof   (d) watertight

2. Metal surface type enclosures in damp or wet locations must be mounted so there is at least _____ airspace between the enclosure and the wall or supporting surface.

   (a) 1/16 in.   (b) 1 1/4 in.   (c) 1/4 in.   (d) 6 in.

3. Where raceways or cables enter above the level of uninsulated live parts of an enclosure in a wet location, a(n) _____ must be used.

   (a) fitting listed for wet locations   (b) explosion proof seal-off
   (c) fitting listed for damp locations   (d) insulated fitting

4. In walls constructed of wood or other _____ material, electrical cabinets must be flush with the finished surface or project therefrom.

   (a) nonconductive   (b) porous   (c) fibrous   (d) combustible

5. Plaster, drywall, or plasterboard surfaces that are broken or incomplete must be repaired so there will be no gaps or open spaces greater than _____ at the edge of a cabinet or cutout box employing a flush-type cover.

   (a) 1/4 in.   (b) 1/2 in.   (c) 1/8 in.   (d) 1/16 in.

6. Openings in cabinets, cutout boxes, and meter socket enclosures through which conductors enter must be _____.

   (a) adequately closed   (b) made using concentric knockouts only
   (c) centered in the cabinet wall   (d) identified

7. Each cable entering a cutout box _____.

   (a) must be secured to the cutout box   (b) can be sleeved through a chase
   (c) must have a maximum of two cables per connector (d) all of these

(• Indicates that 75% or fewer exam takers get the question correct)

8. Cables with entirely nonmetallic sheaths are permitted to enter the top of a surface-mounted enclosure through one or more nonflexible raceways not less than 18 in. or more than _____ ft in length if all of the required conditions are met.

   (a) 3   (b) 10   (c) 25   (d) 100

9. A switch enclosure (cabinet) must not be used as a junction box, except where adequate space is provided so that the conductors don't fill the wiring space at any cross-section to more than 40 percent of the cross-sectional area of the space, and so that _____ don't fill the wiring space at any cross-section to more than 75 percent of the cross-sectional area of the space.

   (a) splices   (b) taps   (c) conductors   (d) all of these

10. For a cabinet or cutout box constructed of sheet steel, the metal must not be thinner than _____ uncoated.

    (a) 0.53 in.   (b) 0.035 in.   (c) 0.053 in.   (d) 1.35 in.

(• Indicates that 75% or fewer exam takers get the question correct)

Article 314 Outlet, Device, Pull and Junction Boxes, Conduit Bodies, and Handhole Enclosures

Article 314 contains installation requirements for outlet boxes, pull and junction boxes, conduit bodies, and handhole enclosures. As with Article 312, conditions of use apply. If you're running a raceway in a hazardous (classified) location, for example, you must use the correct fittings and the proper installation methods. But consider something as simple as a splice. It makes sense that you wouldn't put a splice in the middle of a raceway—doing so means you cannot get to it. But if you put a splice in a conduit body, you're fine, right? Not necessarily. Suppose that conduit body is a "short radius" version (think of it as an elbow with the bend chopped off). You do not have much room inside such an enclosure and for that reason, you cannot put a splice inside a short radius conduit body.

Properly applying Article 314 means you will need to account for the internal volume of all boxes and fittings, and then determine the maximum wire fill. You'll also need to understand many other requirements, which we cover in the *Understanding the NEC, Volume 1* textbook. If you start to get confused, take a break. Look carefully at the illustrations in the accompanying textbook, and you'll learn more quickly and with greater retention.

Questions

1. •Nonmetallic boxes are permitted for use with _____.

    (a) flexible nonmetallic conduit
    (b) liquidtight nonmetallic conduit
    (c) nonmetallic cables and raceways
    (d) all of these

2. Metallic boxes are required to be _____.

    (a) metric (b) installed (c) grounded (d) all of these

3. Short-radius conduit bodies such as capped elbows, and service-entrance elbows that enclose conductors 6 AWG and smaller are intended to enable the installation of the raceway and the contained conductors and must not contain _____.

    (a) splices (b) taps (c) devices (d) any of these

4. Boxes, conduit bodies, and fittings installed in wet locations do not need to be listed for use in wet locations.

    (a) True (b) False

5. According to the *NEC*, the volume of a 3 x 2 x 2 in. device box is _____

    (a) 12 cu in. (b) 14 cu in. (c) 10 cu in. (d) 8 cu in.

6. When counting the number of conductors in a box, a conductor running through the box with no loop in it is counted as _____ conductor(s).

    (a) one (b) two (c) zero (d) none of these

7. When counting the number of conductors in a box, a conductor running through the box with an unbroken loop not less than twice the minimum length required for free conductors in 300.14 is counted as _____ conductor(s).

    (a) one (b) two (c) zero (d) none of these

(• Indicates that 75% or fewer exam takers get the question correct)

8. Equipment grounding conductor(s), and not more than _____ fixture wires (smaller than 14 AWG) can be omitted from the calculations where they enter the box from a domed luminaire or similar canopy and terminate within that box.

   (a) 2 (b) 3 (c) 4 (d) none of these

9. When determining the number of conductors in a box, and one or more factory or field-supplied internal cable clamps are present in the box, a double volume allowance for the clamps, in accordance with Table 314.16(B), must be made based on the largest conductor present in the box.

   (a) True (b) False

10. •For box fill calculations, a reduction of _____ conductor(s) can be made for one hickey and two internal clamps.

    (a) 1 (b) 2 (c) 3 (d) zero

11. Each yoke or strap containing one or more devices or equipment counts as _____ conductor(s), based on the largest conductor that terminates on that device.

    (a) 1 (b) 2 (c) 3 (d) none

12. •What is the total volume, in cubic inches, for box fill calculations for two internal cable clamps, six 12 THHN conductors, and one single-pole switch?

    (a) 2.00 cu in. (b) 4.50 cu in. (c) 14.50 cu in. (d) 20.25 cu in.

13. When a box contains three equipment grounding conductors that originated outside the box, the three grounding conductors are counted as _____ conductor(s) when determining the number of conductors in a box for box fill calculations.

    (a) 3 (b) 6 (c) 1 (d) 0

14. Splices and taps can be made in conduit bodies that are durably and legibly marked by the manufacturer with their volume and the maximum number of conductors as computed in accordance with Table 314.16(B)

    (a) True (b) False

15. When NM cable is used with nonmetallic boxes no larger than 2¼ x 4 in., securing the cable to the box is not required if the cable is fastened within _____ of that box.

    (a) 6 in. (b) 8 in. (c) 10 in. (d) 12 in.

16. •In noncombustible walls or ceilings, the front edge of a box, plaster ring, extension ring, or listed extender may be set back not more than _____ from the finished surface.

    (a) ⅜ in. (b) ⅛ in. (c) ½ in. (d) ¼ in.

17. In walls or ceilings constructed of wood or other combustible surface material, boxes, plaster rings, extension rings, or listed extenders must _____.

    (a) be flush with the surface
    (b) project from the surface
    (c) a or b
    (d) be set back no more than ¼ in.

18. Plaster, drywall, or plasterboard surfaces that are broken or incomplete around boxes employing a flush-type cover or faceplate must be repaired so there will be no gaps or open spaces larger than _____ at the edge of the box.

    (a) ⅜ in. (b) ¼ in. (c) ⅛ in. (d) ¹⁄₁₆ in.

(• Indicates that 75% or fewer exam takers get the question correct)

19. Surface extensions from a flush-mounted box must be made by mounting and mechanically securing an extension ring over the flush box.

    (a) True   (b) False

20. •Only a _____ wiring method can be used for a surface extension from a cover, and the wiring method must include an equipment grounding conductor.

    (a) solid   (b) flexible   (c) rigid   (d) cord

21. Surface mounted enclosures (boxes) must be _____ the building surface.

    (a) rigidly and securely fastened to
    (b) supported by cables that protrude from
    (c) supported by cable entries from the top and allowed to rest against
    (d) none of these

22. Nails or screws can fasten boxes to structural members of a building using brackets on the outside of the enclosure, or they can pass through the interior within _____ of the back or ends of the enclosure. Screws are not permitted to pass through the box unless exposed threads in the box are protected using approved means to avoid abrasions of conductor insulation.

    (a) ⅛ in.   (b) 1⁄16 in.   (c) ¼ in.   (d) ½ in.

23. A wood brace that is used for mounting a box must have a cross-section not less than nominal _____.

    (a) 1 x 2 in.   (b) 2 x 2 in.   (c) 2 x 3 in.   (d) 2 x 4 in.

24. When mounting an enclosure in a finished surface, the enclosure must be _____ secured to the surface by clamps, anchors, or fittings identified for the application.

    (a) temporarily   (b) partially   (c) never   (d) rigidly

25. Outlet boxes can be secured to suspended-ceiling framing members by mechanical means such as _____, or other means identified for the suspended-ceiling framing member(s).

    (a) bolts   (b) screws   (c) rivets   (d) all of these

26. Outlet boxes can be secured to independent support wires, which are taut and secured at both ends, if the box is supported to the independent support wires using methods identified for the purpose.

    (a) True   (b) False

27. Enclosures not over _____ in size, having threaded entries and that do not contain a device(s) or support a luminaire(s) or other equipment, is considered to be adequately supported where two or more conduits are threaded wrenchtight into the enclosure and each conduit secured within 3 ft.

    (a) 50 cu in.   (b) 75 cu in.   (c) 100 cu in.   (d) 125 cu in.

28. •Enclosures not over 100 cu in. that have threaded entries that support luminaires or contain devices are considered adequately supported where two or more conduits are threaded wrenchtight into the enclosure where each conduit is supported within _____ of the enclosure.

    (a) 12 in.   (b) 18 in.   (c) 24 in.   (d) 30 in.

(• Indicates that 75% or fewer exam takers get the question correct)

29. Boxes can be supported from a multiconductor cord or cable, provided the conductors are protected from _____.

    (a) strain   (b) temperature   (c) sunlight   (d) abrasion

30. In completed installations, each outlet box must have a _____.

    (a) cover   (b) faceplate   (c) canopy   (d) any of these

31. Outlet boxes used at luminaire or lamp holder outlets must be _____.

    (a) designed for the purpose   (b) metal only
    (c) plastic only   (d) mounted using bar hangers only

32. A wall-mounted luminaire weighing not more than _____ can be supported to a device box with no fewer than two No. 6 or larger screws.

    (a) 4 lbs   (b) 6 lbs   (c) 8 lbs   (d) 10 lbs

33. Luminaires must be supported independently of the outlet box where the weight exceeds _____

    (a) 60 lbs   (b) 50 lbs   (c) 40 lbs   (d) 30 lbs

34. A luminaire that weighs more than 50 lbs is permitted to be supported by an outlet box or fitting that is designed and listed for the weight of the luminaire.

    (a) True   (b) False

35. When installing floor boxes, boxes _____ must be used.

    (a) made only of metal   (b) listed specifically for this application
    (c) fed by metal raceways only   (d) fed by nonmetallic cable only

36. Where a box is used as the sole support of a ceiling-suspended (paddle) fan, the box must be listed for the application and must be marked with the weight of the fan to be supported if over 35 lbs.

    (a) True   (b) False

37. Listed outlet boxes, or outlet box systems that are identified for the purpose are permitted to support ceiling-suspended fans that weigh more than 35 lbs but no more than _____ if the allowable weight is marked on the box.

    (a) 50 lbs   (b) 60 lbs   (c) 70 lbs   (d) none of these

38. When sizing a pull box in a straight run which contains conductors of 4 AWG or larger, the length of the box must not be less than _____ for systems not over 600V.

    (a) 8 times the diameter of the largest raceway
    (b) 6 times the diameter of the largest raceway
    (c) 48 times the outside diameter of the largest shielded conductor
    (d) 36 times the largest conductor

39. The wiring contained inside which of the following are required to be accessible?

    (a) Outlet boxes   (b) Junction boxes   (c) Pull boxes   (d) all of these

(• Indicates that 75% or fewer exam takers get the question correct)

40. Listed boxes designed for underground installation can be directly buried when covered by _____ if their location is identified and accessible.

    (a) concrete
    (b) gravel
    (c) noncohesive granulated soil
    (d) b or c

41. Handhole enclosures must be designed and installed to withstand _____.

    (a) 3,000 lbs (b) 6,000 lbs (c) all loads likely to be imposed (d) 600 lbs

42. Handhole enclosures must be sized in accordance with 314.28(A) for conductors operating at 600 volts and below. For handhole enclosures without bottoms, the measurement to the removable cover is taken from the _____.

    (a) end of the conduit or cable assembly
    (b) lowest point in the hole
    (c) leveling marks provided
    (d) highest possible ground water level

43. Underground raceways and cable assemblies entering a handhole enclosure must extend into the enclosure, but they are not required to be _____.

    (a) bonded
    (b) insulated
    (c) mechanically connected to the handhole enclosure
    (d) below minimum cover requirements after leaving the handhole

44. Where handhole enclosures without bottoms are installed, all enclosed conductors and any splices or terminations, if present, must be listed as _____.

    (a) suitable for wet locations
    (b) suitable for damp locations
    (c) handhole ready
    (d) general duty

45. Handhole enclosure covers must have an identifying _____ that prominently identifies the function of the enclosure, such as "electric."

    (a) mark (b) logo (c) a or b (d) manual

46. Handhole enclosure covers must require the use of tools to open, or they must weigh over _____. Metal covers and other exposed conductive surfaces must be bonded to an effective ground-fault current path.

    (a) 45 lbs (b) 100 lbs (c) 70 lbs (d) 200 lbs

(• Indicates that 75% or fewer exam takers get the question correct)

Article 320 Armored Cable (Type AC)

Article Overview

Armored cable is an assembly of insulated conductors, 14 AWG through 1 AWG, that are individually wrapped within waxed paper and contained within a flexible spiral metal sheath. Armored cable looks like flexible metal conduit.

Questions

1. •The use of Type AC cable is permitted in _____ installations.

   (a) wet   (b) cable tray   (c) exposed   (d) b and c

2. Armored cable is limited or not permitted _____.

   (a) in damp or wet locations
   (b) where subject to physical damage
   (c) where exposed to corrosive fumes or vapors
   (d) all of these

3. Exposed runs of Type AC cable must closely follow the surface of the building finish or of running boards. Exposed runs are also permitted to be installed on the underside of joists where supported at each joist and located so as not to be subject to physical damage.

   (a) True   (b) False

4. Type AC cable installed through, or parallel to, framing members must be protected against physical damage from penetration by screws or nails.

   (a) True   (b) False

5. Where run across the top of floor joists, within 7 ft of floor or floor joists, or across the face of rafters or studding in attics and roof spaces that are accessible by permanent stairs or ladders, Type AC cable must be protected by substantial guard strips that are _____.

   (a) at least as high as the cable   (b) constructed of metal   (c) made for the cable   (d) none of these

6. When Type AC cable is run across the top of a floor joist in an attic without permanent ladders or stairs, substantial guard strips within _____ of the scuttle hole, or attic entrance, must protect the cable.

   (a) 7 ft   (b) 6 ft   (c) 5 ft   (d) 3 ft

7. When armored cable is run parallel to the sides of rafters, studs, or floor joists in an accessible attic, the cable must be protected with running boards.

   (a) True   (b) False

8. The radius of the curve of the inner edge of any bend must not be less than _____ for AC cable.

   (a) five times the largest conductor within the cable
   (b) three times the diameter of the cable
   (c) five times the diameter of the cable
   (d) six times the outside diameter of the conductors

(• Indicates that 75% or fewer exam takers get the question correct)

Mike Holt Enterprises, Inc. • www.NECcode.com • 1.888.NEC.Code

9. Type AC cable must be supported and secured at intervals not exceeding 4½ ft and the cable must be secured within _____ of every outlet box, cabinet, conduit body, or other armored cable termination.

   (a) 4 in.　　　　(b) 8 in.　　　　(c) 9 in.　　　　(d) 12 in.

10. Armored cable used for the connection of recessed luminaires or equipment within an accessible ceiling does not need to be secured for lengths up to _____.

    (a) 2 ft　　　　(b) 3 ft　　　　(c) 4 ft　　　　(d) 6 ft

11. At all Type AC cable terminations, a(n) _____ must be provided.

    (a) fitting (or box design) that protects the wires from abrasion
    (b) insulating bushing between the conductors and the cable armor
    (c) both a and b
    (d) none of these

12. When Type AC cable is installed in thermal insulation, it must have conductors that are rated at 90°C. The ampacity of the cable in this application is _____.

    (a) based on 90°C column　　(b) as labeled by the manufacturer
    (c) based on the 60°C column　(d) none of these

13. Type AC cable must provide _____ for equipment grounding as required by Article 250.

    (a) an adequate path
    (b) a green terminal on all Type AC fittings
    (c) a solid copper insulated green conductor in all Type AC cables
    (d) a bonding locknut on all Type AC fittings

(• Indicates that 75% or fewer exam takers get the question correct)

Article 330 Metal-Clad Cable (Type MC)

Article Overview

Metal-clad cable encloses one or more insulated conductors in a metal sheath of either corrugated or smooth copper or aluminum tubing, or spiral interlocked steel or aluminum. The physical characteristics of MC cable make it a versatile wiring method you can use in almost any location and for almost any application. The most common type of MC cable is the interlocking type, which looks similar to armored cable or flexible metal conduit. Because the outer sheath of interlocking type cable isn't suitable as an effective ground-fault current path, it contains a separate equipment grounding (bonding) conductor.

Questions

1. Type MC cable must not be used where exposed to the following destructive corrosive condition(s), unless the metallic sheath is suitable for the condition(s) or is protected by material suitable for the condition(s):

   (a) Direct burial in the earth   (b) In concrete   (c) In cinder fill   (d) all of these

2. Type MC cable installed through, or parallel to, framing members must be protected against physical damage from penetration by screws or nails by 1¼ in. separation or protected by a suitable metal plate.

   (a) True   (b) False

3. Type MC cable installed in accessible attics or roof spaces must comply with the same requirements as given for AC cable in 320.24. This includes the installation of _____ to protect the cable when run across the top of floor joists if the space is accessible by permanent stairs or ladders.

   (a) GFCI protection   (b) arc-fault protection   (c) rigid metal conduit   (d) guard strips

4. Type MC cable installed in accessible attics or roof spaces must comply with the same requirements as given for AC cable in 320.24. This includes the installation of guard strips to protect the cable when run across the top of floor joists within _____ of the nearest edge of the scuttle hole or attic entrance if the space is not accessible by permanent stairs or ladders.

   (a) 6 ft   (b) 7 ft   (c) 1¼ in.   (d) 18 in.

5. Smooth-sheath Type MC cable with an external diameter of not greater than 1 in. must have a bending radius of not more than _____ times the cable external diameter.

   (a) 5   (b) 10   (c) 12   (d) 13

6. Bends made in interlocked or corrugated sheath metal clad cable must maintain a bending radius of at least _____ the external diameter of the metallic sheath.

   (a) 5 times   (b) 7 times   (c) 10 times   (d) 125% of

7. Type MC cable must be supported and secured at intervals not exceeding _____.

   (a) 3 ft   (b) 6 ft   (c) 4 ft   (d) 2 ft

(• Indicates that 75% or fewer exam takers get the question correct)

8. Type MC cable containing four or fewer conductors, sized no larger than 10 AWG, must be secured within _____ of every box, cabinet, fitting, or other cable termination.

   (a) 8 in. (b) 18 in. (c) 12 in. (d) 24 in.

9. Type MC cable can be unsupported where it is:

   (a) Fished between concealed access points in finished buildings or structures and support is impracticable.
   (b) Not more than 2 ft in length at terminals where flexibility is necessary.
   (c) Not more than 6 ft from the last point of support within an accessible ceiling for the connection of luminaires.
   (d) a or c

10. Fittings used for connecting Type MC cable to boxes, cabinets, or other equipment must _____.

    (a) be nonmetallic only
    (b) be listed and identified for such use
    (c) be listed and identified as weatherproof
    (d) include anti-shorting bushings (red heads)

11. Where MC cable is used for equipment grounding it must comply with 250.118(10) which allows the metallic sheath alone of interlocked metal tape-type MC cable to be used as an equipment grounding conductor.

    (a) True (b) False

(• Indicates that 75% or fewer exam takers get the question correct)

Article 334 Nonmetallic-Sheathed Cable (Types NM and NMC)

Article Overview

Nonmetallic-sheathed cable is flexible, inexpensive, and easily installed. It provides very limited physical protection of the conductors, so the installation restrictions are strict. However, its low cost and relative ease of installation make it a common wiring method used for residential and commercial branch circuits.

Questions

1. Type NM cable must be _____.

   (a) marked (b) approved (c) identified (d) listed

2. Types NM and NMC nonmetallic-sheathed cables can be used in _____ except as prohibited in 334.12.

   (a) one-family dwellings
   (b) multifamily dwellings of Types III, IV, and V construction
   (c) two family dwellings
   (d) all of these

3. Type NM and Type NMC cables must not be used in one- and two-family dwellings exceeding three floors above grade.

   (a) True (b) False

4. Type NM cable can be installed in multifamily dwellings of Types III, IV, and V construction except as prohibited in 334.12.

   (a) True (b) False

5. Type NM cable can be installed as open runs in dropped or suspended ceilings in other than one- and two-family and multifamily dwellings.

   (a) True (b) False

6. Type NM cable must not be used _____.

   (a) in commercial buildings
   (b) in the air void of masonry block not subject to excessive moisture
   (c) for exposed work
   (d) embedded in poured cement, concrete, or aggregate

7. Type NM cable must closely follow the surface of the building finish or running boards when run exposed.

   (a) True (b) False

8. When installed in _____, nonmetallic-sheathed cable must be protected from physical damage where necessary by RMC, IMC, Schedule 80 rigid nonmetallic conduit, EMT, guard strips, or other means.

   (a) hazardous locations of commercial garages
   (b) exposed work
   (c) service entrance applications
   (d) motion picture studios

(• Indicates that 75% or fewer exam takers get the question correct)

9. Where Type NMC cable is run at angles with joists in unfinished basements, it is permissible to secure cables not smaller than _____ conductors directly to the lower edges of the joist.

   (a) two, 6 AWG (b) three, 8 AWG (c) three, 10 AWG (d) a or b

10. NM cable on a wall of an unfinished basement is permitted to be installed in a listed raceway. A _____ must be installed at the point where the cable enters the raceway. Metal conduit and tubing and metal outlet boxes must be grounded.

    (a) nonmetallic bushing or adapter (b) sealing fitting
    (c) bonding bushing (d) junction box

11. Type NM cable installed through, or parallel to, framing members must be protected against physical damage in accordance with 300.4. Grommets or bushings for the protection of Type NM cable as required in 300.4(B)(1) must be _____ for the purpose, and they must remain in place.

    (a) marked (b) approved (c) identified (d) listed

12. Nonmetallic-sheathed cable installed in accessible attics or roof spaces must comply with the same requirements as given for AC cable in 320.24. This includes the installation of _____ to protect the cable when run across the top of floor joists if the space is accessible by permanent stairs or ladders.

    (a) GFCI protection (b) Arc-fault protection (c) rigid metal conduit (d) guard strips

13. Nonmetallic-sheathed cable installed in accessible attics or roof spaces must comply with the same requirements as given for AC cable in 320.24. This includes the installation of guard strips to protect the cable when run across the top of floor joists within _____ of the nearest edge of the scuttle hole or attic entrance if the space is not accessible by permanent stairs or ladders.

    (a) 6 ft (b) 7 ft (c) 1¼ in. (d) 18 in.

14. Bends made in nonmetallic-sheathed cable must be made so that the cable will not be damaged. The radius of the curve of the inner edge of any bend during or after installation must not be less than _____ the external diameter of the cable.

    (a) 5 times (b) 7 times (c) 10 times (d) 125% of

15. Type NM cable must be secured in place within _____ of every cabinet, box, or fitting.

    (a) 6 in. (b) 10 in. (c) 12 in. (d) 18 in.

16. Flat two-conductor Type NM cables cannot be stapled on edge.

    (a) True (b) False

17. Sections of Type NM cable protected from physical damage by a raceway are not required to be _____ within the raceway.

    (a) covered (b) insulated (c) secured (d) unspliced

18. Type NM cables run horizontally through framing are considered supported and secured where such support does not exceed 4 1/2 ft intervals and the Type NM cable is securely fastened in place within 12 in. of each box, cabinet, or conduit body.

    (a) True (b) False

19. Type NM cable, installed within accessible ceilings for the connections to luminaires and equipment, does not need to be secured within 12 in. from the luminaire or equipment when the free length does not exceed _____.

    (a) 4½ ft (b) 2½ ft (c) 3½ ft (d) any of these

(• Indicates that 75% or fewer exam takers get the question correct)

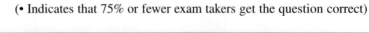

20. The ampacity of Type NM cable must be that of 60°C conductors, as listed in 310.15. However, the 90°C rating can be used for ampacity derating purposes provided the final derated ampacity does not exceed that of a _____ rated conductor.

    (a) 120°C   (b) 60°C   (c) 90°C   (d) none of these

21. Where more than two NM cables containing two or more current-carrying conductors are bundled together and pass through wood framing that is to be fire- or draft-stopped using thermal insulation or sealing foam, the allowable ampacity of each conductor is _____.

    (a) no more than 20A   (b) adjusted in accordance with 310.15(B)(2)(a)
    (c) limited to 30A   (d) calculated by an engineer

22. The insulation rating of ungrounded conductors in Type NM cable must be _____.

    (a) 60°C   (b) 75°C   (c) 90°C   (d) any of these

23. •The difference in the construction specifications between Type NM cable and Type NMC cable is that Type NMC cable is _____ which Type NM is not.

    (a) corrosion-resistant   (b) flame-retardant   (c) fungus-resistant   (d) a and c

(• Indicates that 75% or fewer exam takers get the question correct)

Article 336 Power and Control Tray Cable (Type TC)

Article Overview

Tray Cable is primarily used for installation in cable trays and associated raceways. It can also be used where supported by a messenger wire. It has an outer sheath that is much stiffer than Type NM cable, particularly when cold, and can be difficult to remove. It is usually labeled as "sunlight resistant," although this is not part of the *Code* requirement for this cable type.

Questions

1. Type TC cable can be used _____.

   (a) for power and lighting circuits
   (b) in cable trays in hazardous locations
   (c) in Class 1 control circuits
   (d) all of these

2. Type TC tray cable must not be installed _____.

   (a) where it will be exposed to physical damage
   (b) outside of a raceway or cable tray system
   (c) direct buried unless identified for such use
   (d) all of these

(• Indicates that 75% or fewer exam takers get the question correct)

Article 338 Service-Entrance Cable (Types SE and USE)

Article Overview

Service-entrance cable is a single conductor or multiconductor assembly with or without an overall moisture-resistant covering. This cable is used primarily for services not over 600V, but can also be used for feeders and branch circuits.

Questions

1. Type _____ is a type of multiconductor cable permitted for use as an underground service-entrance cable.

   (a) SE   (b) NMC   (c) UF   (d) USE

2. Type USE cable used for service laterals can emerge from the ground outside at termination in meter bases or other enclosures where protected in accordance with 300.5(D).

   (a) True   (b) False

3. Type SE service entrance cable is permitted for use as _____ in wiring systems where all of the circuit conductors of the cable are of the rubber-covered or thermoplastic type.

   (a) branch circuits   (b) feeders   (c) a or b   (d) neither a or b

4. Type SE service-entrance cables are permitted for use for branch circuits or feeders where the insulated conductors are used for circuit wiring and the uninsulated conductor is used only for _____ purposes.

   (a) grounded neutral connection   (b) equipment grounding
   (c) remote control and signaling   (d) none of these

5. Type SE service entrance cable may be used for interior wiring as long as it complies with the installation requirements of Parts I and II of Article 334, excluding 334.80.

   (a) True   (b) False

6. Bends made in USE and SE cable must be made so that the cable is not damaged. The radius of the curve of the inner edge of any bend during or after installation must not be less than _____ the diameter of the cable.

   (a) 5 times   (b) 7 times   (c) 10 times   (d) 125% of

7. •Type USE or SE cable must have a minimum of _____ conductors (including the uninsulated one) in order for one of the conductors to be uninsulated.

   (a) one   (b) two   (c) three   (d) four

(• Indicates that 75% or fewer exam takers get the question correct)

Article 340 Underground Feeder and Branch-Circuit Cable (Type UF)

Article Overview

UF cable is a moisture-, fungus-, and corrosion-resistant cable system suitable for direct burial in the earth.

Questions

1. Type UF cable must not be used where subject to physical damage. When this cable is subject to physical damage, it must be protected by a suitable method as described in 300.5.

   (a) True   (b) False

2. Underground feeder and branch circuit (Type UF) cable is allowed to be used as service entrance cable.

   (a) True   (b) False

3. Underground Feeder and branch circuit (Type UF) cable is allowed to be used in commercial garages.

   (a) True   (b) False

4. Type UF cable must not be used _____.

   (a) in any hazardous (classified) location
   (b) embedded in poured cement, concrete, or aggregate
   (c) where exposed to direct rays of the sun, unless identified as sunlight-resistant
   (d) all of these

5. Bends made in UF cable must be made so that the cable is not damaged. The radius of the curve of the inner edge of any bend during or after installation must not be less than _____ the diameter of the cable.

   (a) 5 times   (b) 7 times   (c) 10 times   (d) 125% of

6. The ampacity of Type UF cable must be that of _____ in accordance with 310.15.

   (a) 90°C conductors   (b) 75°C conductors   (c) 60°C conductors   (d) none of these

7. The maximum size of conductors in Underground Feeder (Type UF) cable is _____ AWG.

   (a) 14   (b) 10   (c) 1/0   (d) 4/0

8. The overall covering of Type UF cable must be _____.

   (a) flame retardant
   (b) moisture, fungus, and corrosion resistant
   (c) suitable for direct burial in the earth
   (d) all of these

(• Indicates that 75% or fewer exam takers get the question correct)

Article 342 Intermediate Metal Conduit (Type IMC)

Article Overview

Intermediate metal conduit (IMC) is a circular metal raceway with an outside diameter equal to that of rigid metal conduit. The wall thickness of intermediate metal conduit is less than that of rigid metal conduit (RMC), so it has a greater interior cross-sectional area. Intermediate metal conduit is lighter and less expensive than rigid metal conduit, but it can be used in all of the same locations as rigid metal conduit. Intermediate metal conduit also uses a different steel alloy that makes it stronger than rigid metal conduit, even though the walls are thinner. IMC is manufactured in both galvanized steel and aluminum; the steel type is much more common.

Questions

1. IMC can be installed in or under cinder fill subject to permanent moisture _____.

   (a) where the conduit is not less than 18 in. under the fill
   (b) when protected on all sides by 2 in. of noncinder concrete
   (c) where protected by corrosion protection judged suitable for the condition
   (d) any of these

2. Materials such as straps, bolts, screws, etc. that are associated with the installation of IMC in wet locations are required to be _____.

   (a) weatherproof (b) weathertight (c) corrosion-resistant (d) none of these

3. Where practicable, contact of dissimilar metals must be avoided anywhere in an IMC raceway installation to prevent _____.

   (a) corrosion (b) galvanic action (c) shorts (d) none of these

4. One-inch IMC raceway containing three or more conductors must not exceed _____ percent conductor fill.

   (a) 53 (b) 31 (c) 40 (d) 60

5. The cross-sectional area of 1 in. IMC is approximately _____

   (a) 1.22 sq in. (b) 0.62 sq in. (c) 0.96 sq in. (d) 2.13 sq in.

6. A run of IMC must not contain more than the equivalent of _____ quarter bends including all offsets between pull points such as conduit bodies and boxes.

   (a) 1 (b) 2 (c) 3 (d) 4

7. IMC must be firmly fastened within _____ of each outlet box, junction box, device box, fitting, cabinet, or other conduit termination.

   (a) 12 in. (b) 18 in. (c) 2 ft (d) 3 ft

8. One-inch IMC must be supported every _____ .

   (a) 8 ft (b) 10 ft (c) 12 ft (d) 14 ft

(• Indicates that 75% or fewer exam takers get the question correct)

9. Exposed vertical risers of IMC for industrial machinery or fixed equipment can be supported at intervals not exceeding _____ if the conduit is made up with threaded couplings, firmly supported at the top and bottom of the riser, and no other means of support is available.

   (a) 10 ft   (b) 12 ft   (c) 15 ft   (d) 20 ft

10. Horizontal runs of IMC supported by openings through framing members at intervals not exceeding 10 ft and securely fastened within 3 ft of terminations is permitted.

    (a) True   (b) False

11. Threadless couplings approved for use with IMC in wet locations must be _____.

    (a) rainproof   (b) listed for wet locations   (c) moistureproof   (d) concrete-tight

12. Threadless couplings and connectors must not be used on threaded IMC ends unless the fittings are listed for the purpose.

    (a) True   (b) False

13. Running threads of IMC must not be used on conduit for connection at couplings.

    (a) True   (b) False

14. Where intermediate metal conduit enters a box, fitting, or other enclosure, _____ must be provided to protect the wire from abrasion.

    (a) a bushing   (b) duct seal   (c) electrical tape   (d) seal off fittings

(• Indicates that 75% or fewer exam takers get the question correct)

Article 344 Rigid Metal Conduit (Type RMC)

Article Overview

Rigid metal conduit (RMC), commonly called "rigid," has long been the standard raceway for providing protection from physical impact and from difficult environments. The outside diameter of rigid metal conduit is the same as intermediate metal conduit. However, the wall thickness of rigid metal conduit is greater than intermediate metal conduit, therefore it has a smaller interior cross-sectional area. Rigid metal conduit is heavier and more expensive than intermediate metal conduit, and it can be used in any location. RMC is manufactured in both galvanized steel and aluminum; the steel type is much more common.

Questions

1. RMC can be installed in concrete, in direct contact with the earth, or in areas subject to severe corrosive influences when protected by _____ and judged suitable for the condition.

    (a) ceramic (b) corrosion protection (c) backfill (d) a natural barrier

2. RMC can be installed in or under cinder fill subject to permanent moisture when protected on all sides by a layer of noncinder concrete not less than _____ thick.

    (a) 2 in. (b) 4 in. (c) 6 in. (d) 18 in.

3. Materials such as straps, bolts, etc., associated with the installation of RMC in a wet location are required to be _____.

    (a) weatherproof (b) weathertight (c) corrosion-resistant (d) none of these

4. Aluminum fittings and enclosures can be used with _____ conduit where not subject to severe corrosive influences.

    (a) steel rigid metal (b) aluminum rigid metal (c) PVC-coated rigid conduit only (d) a and b

5. The minimum radius for a field bend of 1 in. rigid metal conduit is _____, when using a one-shot bender.

    (a) 10½ in. (b) 11½ in. (c) 5¾ in. (d) 9½ in.

6. The minimum radius of a field bend on 1¼ in. RMC is _____.

    (a) 7 in. (b) 8 in. (c) 14 in. (d) 10 in.

7. All cut ends of rigid metal conduit must be _____ or otherwise finished to remove rough edges.

    (a) threaded (b) reamed (c) painted (d) galvanized

8. When rigid metal conduit is threaded in the field, a standard die with _____ must be used.

    (a) ¾ in. taper per foot (b) 1 in. taper per foot (c) 1/16 in. taper per foot (d) no taper

9. Two-inch RMC must typically be supported every _____.

    (a) 10 ft (b) 12 ft (c) 14 ft (d) 15 ft

(• Indicates that 75% or fewer exam takers get the question correct)

10. Straight runs of 1 in. RMC using threaded couplings may be secured at intervals not exceeding _____.

    (a) 5 ft   (b) 10 ft   (c) 12 ft   (d) 14 ft

11. Exposed vertical risers of RMC for industrial machinery or fixed equipment can be supported at intervals not exceeding _____ if the conduit is made up with threaded couplings, firmly supported at the top and bottom of the riser, and no other means of support is available.

    (a) 6 ft   (b) 10 ft   (c) 20 ft   (d) none of these

12. Horizontal runs of RMC supported by openings through _____ at intervals not exceeding 10 ft and securely fastened within 3 ft of termination points are permitted.

    (a) walls   (b) trusses   (c) rafters   (d) framing members

13. When threadless couplings and connectors used in the installation of RMC are buried in masonry or concrete, they must be of the _____ type.

    (a) raintight   (b) wet and damp location   (c) nonabsorbent   (d) concrete-tight

14. Threadless couplings and connectors used with RMC and installed in wet locations must be _____.

    (a) listed for wet locations   (b) listed for damp locations   (c) nonabsorbent   (d) weatherproof

15. Running threads must not be used on rigid metal conduit for connection at _____.

    (a) boxes   (b) cabinets   (c) couplings   (d) meter sockets

16. Where rigid metal conduit enters a box, fitting, or other enclosure, a bushing must be provided to protect the wire from abrasion unless the design of the box, fitting, or enclosure is such as to afford equivalent protection.

    (a) True   (b) False

(• Indicates that 75% or fewer exam takers get the question correct)

Article 348 Flexible Metal Conduit (Type FMC)

Article Overview

Flexible metal conduit (FMC), commonly called Greenfield or "flex," is a raceway of an interlocked metal strip of either steel or aluminum. It's primarily used for the final 6 ft or less of raceways between a more rigid raceway system and equipment that moves, shakes, or vibrates. Examples of such equipment include pump motors and industrial machinery.

Questions

1. Which of the following conductor types are required to be used when FMC is installed in a wet location?

    (a) THWN (b) XHHW (c) THW (d) any of these

2. FMC cannot be installed _____.

    (a) underground
    (b) embedded in poured concrete
    (c) where subject to physical damage
    (d) all of these

3. FMC can be installed exposed or concealed where not subject to physical damage.

    (a) True (b) False

4. The largest size THHN conductor permitted in ⅜ in. FMC is _____ AWG.

    (a) 12 (b) 16 (c) 14 (d) 10

5. How many 12 AWG XHHW conductors, not counting a bare ground wire, are allowed in trade size ⅜ FMC (maximum of 6 ft) with outside fittings?

    (a) 4 (b) 3 (c) 2 (d) 5

6. Bends in flexible metal conduit must be made so that the conduit is not damaged and the internal diameter of the conduit is _____. The radius of the curve to the centerline of any bend must not be less than shown in Table 2, Chapter 9 using the column for "Other Bends."

    (a) larger than ⅜ in. (b) not effectively reduced (c) increased (d) larger than 1 in.

7. Bends in flexible metal conduit _____ between pull points.

    (a) must not be made
    (b) need not be limited (in degrees)
    (c) must not exceed 360 degrees
    (d) must not exceed 180 degrees

8. All cut ends of flexible metal conduit must be trimmed or otherwise finished to remove rough edges, except where fittings _____.

    (a) are the crimp-on type
    (b) thread into the convolutions
    (c) contain insulated throats
    (d) are listed for grounding

(• Indicates that 75% or fewer exam takers get the question correct)

9. FMC must be supported and secured _____.

   (a) at intervals not exceeding 4½ ft
   (b) within 8 in. on each side of a box where fished
   (c) where fished
   (d) at intervals not exceeding 6 ft at motor terminals

10. Unsupported lengths of flexible metal conduit are allowed at terminals where flexibility is required but must not exceed _____.

    (a) 3 ft for trade sizes ½ in. through 1¼ in.
    (b) 4 ft for trade sizes 1½ in. through 2 in.
    (c) 5 ft for trade size 2½ in. and larger
    (d) all of these

11. Horizontal runs of flexible metal conduit supported by openings through framing members at intervals not greater than _____ and securely fastened within 12 in. of termination points are permitted.

    (a) 1.4 ft   (b) 12 in.   (c) 4½ ft   (d) 6 ft

12. In a concealed FMC installation, _____ connectors must not be used.

    (a) straight   (b) angle   (c) grounding-type   (d) none of these

13. When flexible metal conduit is used to install equipment where flexibility is required, _____ must be installed.

    (a) an equipment grounding conductor
    (b) an expansion fitting
    (c) flexible nonmetallic connectors
    (d) a grounded neutral conductor one size larger

(• Indicates that 75% or fewer exam takers get the question correct)

Article 350 Liquidtight Flexible Metal Conduit (Type LFMC)

Article Overview

Liquidtight flexible metal conduit (LFMC), with its associated connectors and fittings, is a flexible raceway system commonly used for connections to equipment that vibrate or are required to move occasionally. Liquidtight flexible metal conduit is commonly called Sealtight® or "liquidtight." Liquidtight flexible metal conduit is of similar construction to flexible metal conduit, but has an outer liquidtight thermoplastic covering. It has the same primary purpose as flexible metal conduit, but it also provides protection from moisture and corrosive effects.

Questions

1. The use of listed and marked LFMC is permitted for _____.

   (a) direct burial where listed and marked for the purpose
   (b) exposed work
   (c) concealed work
   (d) all of these

2. •LFMC smaller than _____ must not be used, except as permitted in 348.20(A).

   (a) ⅜ in.   (b) ½ in.   (c) 1½ in.   (d) 1¼ in.

3. The maximum number of 14 THHN conductors permitted in ⅜ in. LFMC with outside fittings is _____.

   (a) 4   (b) 7   (c) 5   (d) 6

4. Where flexibility is necessary, securing LFMC is not required for lengths not exceeding _____ at terminals.

   (a) 2 ft   (b) 3 ft   (c) 4 ft   (d) 6 ft

5. When LFMC is used as a fixed raceway, it must be secured within _____ in. on each side of the box and must be supported and secured at intervals not exceeding _____ ft.

   (a) 12, 4½   (b) 18, 3   (c) 12, 3   (d) 18, 4

6. Liquidtight flexible metal conduit is not required to be fastened when used for tap conductors to luminaires up to _____ in length.

   (a) 4½ ft   (b) 18 in.   (c) 6 ft   (d) no limit on length

7. Horizontal runs of liquidtight flexible metal conduit supported by openings through framing members at intervals not greater than _____ and securely fastened within 12 in. of termination points are permitted.

   (a) 1.4 ft   (b) 12 in.   (c) 4½ ft   (d) 6 ft

8. _____ connectors must not be used for concealed installations of liquidtight flexible metal conduit.

   (a) Straight   (b) Angle   (c) Grounding-type   (d) none of these

9. When LFMC is used to connect equipment requiring flexibility, a separate _____ conductor must be installed.

   (a) main bonding jumper   (b) grounded   (c) equipment grounding   (d) none of these

(• Indicates that 75% or fewer exam takers get the question correct)

10. Where flexibility _____ liquidtight flexible metal conduit is permitted to be used as an equipment grounding conductor when installed in accordance with 250.118(6).

    (a) is required   (b) is not required   (c) either a or d   (d) is optional

(• Indicates that 75% or fewer exam takers get the question correct)

Article 352 Rigid Nonmetallic Conduit (Type RNC)

Article Overview

Rigid nonmetallic conduit (RNC) is commonly called "PVC." This type of conduit gives you many of the advantages of rigid conduit, while allowing installation in areas that are wet or corrosive. It is an inexpensive raceway, and easily installed. It is lightweight, easily cut, glued together, and relatively strong. However, RNC is brittle when cold, and it sags when hot. It is commonly used as an underground raceway because of its low cost, ease of installation, and resistance to corrosion and decay.

Questions

1. Extreme _____ may cause rigid nonmetallic conduit to become brittle, and therefore more susceptible to damage from physical contact.

    (a) sunlight   (b) corrosive conditions   (c) heat   (d) cold

2. Rigid nonmetallic conduit and fittings can be used in areas of dairies, laundries, canneries, or other wet locations and in locations where walls are frequently washed. However, the entire conduit system including boxes and _____ must be installed and equipped to prevent water from entering the conduit.

    (a) luminaires   (b) fittings   (c) supports   (d) all of these

3. Rigid nonmetallic conduit is permitted for exposed work in buildings _____, where not subject to physical damage and if identified for such use.

    (a) three floors and less   (b) twelve floors and less   (c) six floors and less   (d) without height limits

4. Rigid nonmetallic conduit can be used to support nonmetallic conduit bodies not larger than the largest raceway, but the conduit bodies must not contain devices, luminaires, or other equipment.

    (a) True   (b) False

5. Among the uses that are NOT permitted for rigid nonmetallic conduit, RNC must not be used _____.

    (a) in hazardous (classified) locations
    (b) for the support of luminaires or other equipment
    (c) where subject to physical damage unless identified for such use
    (d) all of these

6. The number of conductors allowed in rigid nonmetallic conduit must not exceed that permitted by the percentage fill specified in _____.

    (a) Table 1, Chapter 9   (b) Table 250.66   (c) Table 310.16   (d) 240.6

7. Bends in rigid nonmetallic conduit must be made so that the conduit is not damaged and the internal diameter of the conduit is not effectively reduced. Field bends must be made only _____.

    (a) by hand forming the bend
    (c) with a truck exhaust pipe
    (b) with bending equipment identified for the purpose
    (d) by use of an open flame torch

(• Indicates that 75% or fewer exam takers get the question correct)

8. Bends in rigid nonmetallic conduit must _____ between pull points.

   (a) not be made
   (b) not be limited in degrees
   (c) be limited to not more than 360 degrees
   (d) be limited to 180 degrees

9. When installing rigid nonmetallic conduit, _____.

   (a) all cut ends must be trimmed inside and outside to remove rough edges
   (b) there must be a support within 2 ft of each box and cabinet
   (c) all joints must be made by an approved method
   (d) a and c

10. Rigid nonmetallic conduit must be securely fastened within _____ of each box.

    (a) 6 in.   (b) 24 in.   (c) 12 in.   (d) 36 in.

11. One-inch rigid nonmetallic conduit must be supported every _____, unless otherwise listed.

    (a) 2 ft   (b) 3 ft   (c) 4 ft   (d) 6 ft

12. Expansion fittings for rigid nonmetallic conduit must be provided to compensate for thermal expansion and contraction when the length change in a straight run between securely mounted boxes, cabinets, elbows, or other conduit terminations is expected to be _____ or greater.

    (a) ¼ in.   (b) ½ in.   (c) 1 in.   (d) none of these

13. Where rigid nonmetallic conduit enters a box, fitting, or other enclosure, a bushing or adapter must be provided to protect the wire from abrasion unless the design of the box, fitting, or enclosure is such as to afford equivalent protection.

    (a) True   (b) False

14. All joints between lengths of rigid nonmetallic conduit, and between conduit and couplings, fittings, and boxes must be made by _____.

    (a) the authority having jurisdiction
    (b) set screw fitting
    (c) an approved method
    (d) expansion fittings

15. An equipment grounding conductor is not required in rigid nonmetallic conduit if the grounded neutral conductor is used to ground equipment as permitted in 250.142.

    (a) True   (b) False

16. Rigid nonmetallic conduit and fittings must be composed of suitable nonmetallic material that is resistant to moisture and chemical atmospheres. For use above ground it must have additional characteristics including _____.

    (a) flame retardance
    (b) resistance to low temperatures and sunlight effects
    (c) resistance to distortion from heat
    (d) all of these

(• Indicates that 75% or fewer exam takers get the question correct)

Article 353 High-Density Polyethylene Conduit (Type HDPE)

Article Overview

This is a new article in the 2005 *NEC*; HDPE was previously included in Article 352. It's lightweight and durable. It resists decomposition, oxidation, and hostile elements that cause damage to other materials. HDPE is mechanically and chemically resistant to a host of environmental conditions. Uses include communications, data, cable television, and general-purpose raceways.

Questions

1. High-density polyethylene conduit (HDPE) can be manufactured _____.

   (a) in discrete lengths
   (b) in continuous lengths from a reel
   (c) only in 20 ft. lengths
   (d) either a or b

2. HDPE is permitted to be installed _____.

   (a) where subject to chemicals for which the conduit is listed
   (b) in cinder fill
   (c) in direct burial installations in earth or concrete
   (d) all of these

3. There is never a case where HDPE can be installed in a hazardous location.

   (a) True (b) False

4. HDPE is not permitted to be installed _____.

   (a) where exposed
   (b) within a building
   (c) for conductors operating at a temperature above the rating of the raceway
   (d) all of these

5. HDPE is not permitted where it will be subject to ambient temperatures in excess of _____.

   (a) 50°C (b) 60°C (c) 75°C (d) 90°C

6. HDPE is allowed only in trade sizes _____.

   (a) ¾ to 4 (b) ½ to 4 (c) 1 to 5 (d) 1 to 3

7. Bends made in HDPE must be made _____.

   (a) in a manner that will not damage the raceway
   (b) so as not to significantly reduce the internal diameter of the raceway
   (c) only with mechanical bending tools
   (d) a and b

8. Bends made in HDPE must not exceed _____ degrees between pull points.

   (a) 180 (b) 270 (c) 360 (d) 480

(• Indicates that 75% or fewer exam takers get the question correct)

Mike Holt Enterprises, Inc. • www.NECcode.com • 1.888.NEC.Code

9. The cut ends of HDPE must be _____ to avoid rough edges.

   (a) filed on the inside
   (b) trimmed inside and outside
   (c) cut only with a hack saw
   (d) all of these

10. When HDPE enters a box, fitting, or other enclosure, a(n) _____ must be provided to protect the wire from abrasion where the box design does not provide such protection.

    (a) bushing
    (b) adapter
    (c) a or b
    (d) reducing bushing

11. Any joints between lengths of HDPE must be made using _____.

    (a) expansion fittings
    (b) an approved method
    (c) a listed method
    (d) none of these

12. Where equipment grounding is required for an installation HDPE, a separate equipment grounding conductor must be _____.

    (a) an insulated copper conductor
    (b) installed within the conduit
    (c) stranded bare copper wire
    (d) a solid bare copper wire

13. HDPE must be resistant to _____.

    (a) moisture
    (b) corrosive chemical atmospheres
    (c) impact and crushing
    (d) all of these

14. Each length of HDPE must be clearly and durably marked not less than every _____ ft, as required in 110.21.

    (a) 10
    (b) 3
    (c) 5
    (d) 20

(• Indicates that 75% or fewer exam takers get the question correct)

Article 354 Nonmetallic Underground Conduit with Conductors (Type NUCC)

Article Overview

Nonmetallic underground conduit with conductors (NUCC) is a factory assembly of conductors or cables inside a nonmetallic, smooth-wall conduit with a circular cross section. The nonmetallic conduit is manufactured from a material called HDPE that is resistant to moisture and corrosive agents. It can also be supplied on reels without damage or distortion and is of sufficient strength to withstand abuse, such as impact or crushing, in handling and during installation without damage to conduit or conductors.

Questions

1. NUCC and its associated fittings must be _____.

   (a) listed (b) approved (c) identified (d) none of these

2. The use of NUCC is permitted _____.

   (a) for direct-burial underground installations
   (b) to be encased or embedded in concrete
   (c) in cinder fill
   (d) all of these

3. NUCC must not be used _____.

   (a) in exposed locations
   (b) inside buildings
   (c) in hazardous (classified) locations
   (d) all of these

4. NUCC larger than _____ must not be used.

   (a) 1 in. (b) 2 in. (c) 3 in. (d) 4 in.

5. Bends in nonmetallic underground conduit with conductors (NUCC) must be _____ so that the conduit will not be damaged and the internal diameter of the conduit will not be effectively reduced.

   (a) manually made
   (b) made only with approved benders
   (c) made with rigid metal conduit bending shoes
   (d) made using an open flame torch

6. Bends in nonmetallic underground conduit with conductors (NUCC) must _____ between termination points.

   (a) not be made
   (b) not be limited in degrees
   (c) be limited to not more than 360 degrees
   (d) be limited to 180 degrees

7. In order to _____ NUCC, the conduit must be trimmed away from the conductors or cables using an approved method that will not damage the conductor or cable insulation or jacket.

   (a) facilitate installing
   (b) enhance the appearance of the installation of
   (c) terminate
   (d) provide safety to the persons installing

(• Indicates that 75% or fewer exam takers get the question correct)

8. Where nonmetal underground conduit with conductors (NUCC) enters a box, fitting, or other enclosure, a bushing or adapter must be provided to protect the conductor or cable from abrasion unless the design of the box, fitting, or enclosure is such as to afford equivalent protection.

    (a) True   (b) False

9. All joints between nonmetallic underground conduit with conductors (NUCC), fittings, and boxes must be made by _____.

    (a) a qualified person   (b) set screw fittings   (c) an approved method   (d) exothermic welding

10. NUCC must be capable of being supplied on reels without damage or _____, and must be of sufficient strength to withstand abuse, such as impact or crushing in handling and during installation, without damage to conduit or conductors.

    (a) distortion   (b) breakage   (c) shattering   (d) all of these

(• Indicates that 75% or fewer exam takers get the question correct)

Article 356 Liquidtight Flexible Nonmetallic Conduit (Type LFNC)

Article Overview

Liquidtight flexible nonmetallic conduit (LFNC) is a listed raceway of circular cross section having an outer liquidtight, nonmetallic, sunlight-resistant jacket over an inner flexible core with associated couplings, connectors, and fittings.

Questions

1. Type LFNC-B can be installed in lengths longer than _____ where secured in accordance with 356.30.

    (a) 2 ft (b) 3 ft (c) 6 ft (d) 10 ft

2. The use of listed and marked LFNC is permitted for _____.

    (a) direct burial where listed and marked for the purpose (b) exposed work
    (c) concealed work (d) all of these

3. The number of conductors allowed in LFNC must not exceed that permitted by the percentage fill specified in _____.

    (a) Table 1, Chapter 9 (b) Table 250.66 (c) Table 310.16 (d) 240.6

4. Bends in LFNC must be made so that the conduit will not be damaged and the internal diameter of the conduit will not be effectively reduced. Bends are permitted to be made only _____.

    (a) manually without auxiliary equipment (b) with bending equipment identified for the purpose
    (c) with any kind of conduit bending tool that will work (d) by use of an open flame torch

5. Bends in LFNC must _____ between pull points.

    (a) not be made (b) not be limited in degrees
    (c) be limited to not more than 360 degrees (d) be limited to 180 degrees

6. When LFNC is installed as an exposed raceway without need of flexibility, it must be securely fastened within _____ on each side of the box and must be fastened at intervals not exceeding _____.

    (a) 12 in., 4½ ft (b) 18 in., 3 ft (c) 12 in., 3 ft (d) 18 in., 4 ft

7. Where flexibility is necessary, securing LFNC is not required for lengths less than _____ at terminals.

    (a) 2 ft (b) 3 ft (c) 4 ft (d) 6 ft

8. When LFNC is used to connect equipment requiring flexibility, a separate _____ must be installed.

    (a) equipment grounding conductor (b) expansion fitting
    (c) flexible nonmetallic connector (d) none of these

(• Indicates that 75% or fewer exam takers get the question correct)

Article 358 Electrical Metallic Tubing (Type EMT)

Article Overview

Electrical metallic tubing (EMT) is a lightweight raceway that is relatively easy to bend, cut, and ream. Because it isn't threaded, all connectors and couplings are of the threadless type and provide quick, easy, and inexpensive installation when compared to other metallic conduit systems. This makes it very popular. EMT is manufactured in both galvanized steel and aluminum; the steel type is much more common.

Questions

1. When EMT is installed in wet locations, all support, bolts, straps, screws, and so forth must be _____.

    (a) of corrosion-resistant materials
    (b) protected against corrosion
    (c) a or b
    (d) nonmetallic materials only

2. EMT must not be used where _____.

    (a) subject to severe physical damage
    (b) protected from corrosion only by enamel
    (c) used for the support of luminaires
    (d) any of these

3. The minimum and maximum size of EMT is _____, except for special installations.

    (a) 5/16 and 3 in.   (b) 3/8 and 4 in.   (c) ½ and 3 in.   (d) ½ and 4 in.

4. A run of EMT between outlet boxes must not exceed _____ offsets close to the boxes.

    (a) 360° plus   (b) 360° total including   (c) four quarter bends plus   (d) 180° total including

5. EMT must not be threaded.

    (a) True   (b) False

6. •EMT must be supported within 3 ft of each coupling.

    (a) True   (b) False

7. Fastening of unbroken lengths of EMT conduit can be increased to a distance of _____ from the termination point where the structural members do not readily permit fastening within 3 ft.

    (a) 10 ft   (b) 5 ft   (c) 4 ft   (d) 25 ft

8. Horizontal runs of EMT supported by openings through framing members at intervals not greater than _____, and securely fastened within 3 ft of termination points, are permitted.

    (a) 1.4 ft   (b) 12 in.   (c) 4½ ft   (d) 10 ft

9. Couplings and connectors used with EMT must be made up _____.

    (a) of metal
    (b) in accordance with industry standards
    (c) tight
    (d) none of these

(• Indicates that 75% or fewer exam takers get the question correct)

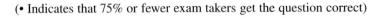

smurf tube ᵈ

Article 362 Electrical Nonmetallic Tubing (Type ENT)

Article Overview

Electrical nonmetallic tubing (ENT) is a pliable, corrugated, circular raceway made of polyvinyl chloride (PVC).

Questions

1. ENT is composed of a material that is resistant to moisture, chemical atmospheres, and is _____.

   (a) flexible   (b) flame-retardant   (c) fireproof   (d) flammable

2. When ENT is installed concealed in walls, floors, and ceilings of buildings exceeding three floors above grade, a thermal barrier must be provided having a minimum _____-minute finish rating as listed for fire-rated assemblies.

   (a) 5   (b) 10   (c) 15   (d) 30

3. When a building is supplied with a(n) _____ fire sprinkler system, ENT can be installed exposed or concealed in buildings of any height.

   (a) listed   (b) identified   (c) NFPA 13-2002 approved   (d) none of these

4. In a building without a fire sprinkler system, ENT is permitted to be installed above a suspended ceiling if the suspended ceiling provides a thermal barrier having at least a _____-minute finish rating as identified in listings of fire-rated assemblies.

   (a) 5   (b) 10   (c) 15   (d) none of these

5. When a building is supplied with an approved fire sprinkler system, ENT is permitted to be installed above any suspended ceiling.

   (a) True   (b) False

6. ENT and fittings can be _____, provided fittings identified for this purpose are used.

   (a) encased in poured concrete
   (b) embedded in a concrete slab on grade where the tubing is placed on sand or approved screenings
   (c) either a or b
   (d) none of these

7. ENT is permitted for direct earth burial when used with fittings listed for this purpose.

   (a) True   (b) False

8. ENT is not permitted in places of assembly unless it is encased in at least _____ of concrete.

   (a) 1 in.   (b) 2 in.   (c) 3 in.   (d) 4 in.

9. ENT must not be used where exposed to the direct rays of the sun, unless identified as _____.

   (a) high-temperature rated   (b) sunlight resistant   (c) schedule 80   (d) never can be

(• Indicates that 75% or fewer exam takers get the question correct)

Mike Holt Enterprises, Inc. • www.NECcode.com • 1.888.NEC.Code

10. The number of conductors allowed in ENT must not exceed that permitted by the percentage fill specified in _____.

   (a) Table 1, Chapter 9   (b) Table 250.66   (c) Table 310.16   (d) 240.6

11. In electrical nonmetallic tubing, the maximum number of bends between pull points cannot exceed _____ degrees, including any offsets.

   (a) 320   (b) 270   (c) 360   (d) unlimited

12. ENT must be securely fastened in place every _____.

   (a) 12 in.   (b) 18 in.   (c) 24 in.   (d) 36 in.

13. Bushings or adapters are required at ENT terminations to protect the conductors from abrasion, unless the box, fitting, or enclosure design provides equivalent protection.

   (a) True   (b) False

14. All joints between lengths of ENT, and between ENT and couplings, fittings, and boxes must be made by _____.

   (a) a qualified person   (b) set screw fittings   (c) an approved method   (d) exothermic welding

15. Where equipment grounding is required by Article 250 for ENT installations, a separate equipment grounding conductor must _____.

   (a) be run outside the raceway using solid copper wire
   (b) be installed in the raceway
   (c) be obtained using a separate driven ground rod
   (d) not be required

(• Indicates that 75% or fewer exam takers get the question correct)

Article 376 Metal Wireways

Article Overview

A metal wireway or "trough" is commonly used where access to the conductors within the raceway is required to make terminations, splices, or taps to several devices at a single location. Its cost precludes its use for other than short distances, except in some commercial or industrial occupancies where the wiring is frequently revised.

Questions

1. Metal wireways are sheet metal troughs with _____ for housing and protecting electric wires and cable.

   (a) removable covers (b) hinged covers (c) a or b (d) none of these

2. Metal wireways can be installed either exposed or concealed under all conditions.

   (a) True (b) False

3. Wireways are permitted for _____.

   (a) exposed work (b) concealed work
   (c) wet locations if listed for the purpose (d) a and c

4. Wireways are permitted to pass transversely through a wall _____. Access to the conductors must be maintained on both sides of the wall.

   (a) if the length passing through the wall is unbroken (b) if the wall is not fire rated
   (c) in hazardous locations (d) if the wall is fire rated

5. Conductors larger than that for which the wireway is designed may be installed in any wireway.

   (a) True (b) False

6. The sum of the cross-sectional areas of all contained conductors at any cross section of a metal wireway must not exceed _____.

   (a) 50 percent (b) 20 percent (c) 25 percent (d) 80 percent

7. The derating factors in 310.15(B)(2)(a) must be applied to a metal wireway only where the number of current-carrying conductors in the wireway exceeds _____.

   (a) 30 (b) 20 (c) 80 (d) 3

8. Where insulated conductors are deflected within a metallic wireway, the wireway must be sized to meet the bending requirements corresponding to one wire per terminal in Table 312.6(A).

   (a) True (b) False

(• Indicates that 75% or fewer exam takers get the question correct)

9. Where a wireway is used as a pull box for insulated conductors 4 AWG or larger, the distance between raceway and cable entries enclosing the same conductor must not be less than that required in 314.28(A)(1) for straight pulls and 314.28(A)(2) for angle pulls.

   (a) True   (b) False

10. Wireways must be supported where run horizontally at each end and at intervals not to exceed _____, or for individual lengths longer than _____ at each end or joint, unless listed for other support intervals.

    (a) 5 ft   (b) 10 ft   (c) 3 ft   (d) 6 ft

11. Vertical runs of metal wireways must be securely supported at intervals not exceeding _____ and must not have more than one joint between supports.

    (a) 5 ft   (b) 20 ft   (c) 10 ft   (d) 15 ft

12. Splices and taps are permitted within a metal wireway provided they are accessible. The conductors, including splices and taps, must not fill the wireway to more than _____ percent of its area at that point.

    (a) 25   (b) 80   (c) 125   (d) 75

13. Power distribution blocks installed in metal wireways must be listed.

    (a) True   (b) False

14. In addition to the wiring space requirement in 356.56(A), the power distribution block must be installed in a metal wireway not smaller than that specified _____.

    (a) by the wireway manufacturer
    (b) by the manufacturer of the power distribution block
    (c) both a and b
    (d) either a or b

15. Power distribution blocks installed in metal wireways must _____.

    (a) allow for sufficient wire-bending space at terminals
    (b) not have exposed live parts after installation
    (c) either a or b
    (d) both a and b

(• Indicates that 75% or fewer exam takers get the question correct)

Article 378 Nonmetallic Wireways

Article Overview

A nonmetallic wireway is a "trough" made from PVC, fiber reinforced plastic (FRP), fiberglass, or similar materials. It is used in harsh environments, but has severe limitations on its use.

Questions

1. Nonmetallic wireways are permitted for _____.

   (a) exposed work
   (b) concealed work
   (c) wet locations if listed for the purpose
   (d) a and c

2. Nonmetallic wireways can pass transversely through a wall _____.

   (a) if the length through the wall is unbroken
   (b) if the wall is not fire rated
   (c) in hazardous (classified) locations
   (d) if the wall is fire rated

3. The sum of the cross-sectional areas of all contained conductors at any cross section of a nonmetallic wireway must not exceed _____.

   (a) 50 percent (b) 20 percent (c) 25 percent (d) 80 percent

4. The derating factors in 310.15(B)(2)(a) apply to a nonmetallic wireway.

   (a) True (b) False

5. Where a nonmetallic wireway is used as a pull box for insulated conductors 4 AWG or larger, the distance between raceway and cable entries enclosing the same conductor must not be less than that required in 314.28(A)(1) for straight pulls and 314.28(A)(2) for angle pulls.

   (a) True (b) False

6. Nonmetallic wireways must be supported where run horizontally at each end and at intervals not to exceed _____ and at each end or joint, unless listed for other support intervals.

   (a) 5 ft (b) 10 ft (c) 3 ft (d) 6 ft

7. Where run vertically, nonmetallic wireways must be securely supported at intervals not exceeding _____, with no more than one joint between supports.

   (a) 5 ft (b) 10 ft (c) 4 ft (d) 6 ft

8. Expansion fittings for nonmetallic wireways must be provided to compensate for thermal expansion and contraction, where the length change is expected to be _____ or greater in a straight run.

   (a) ¼ in. (b) ½ in. (c) 6 in. (d) ¹⁄₁₆ in.

(• Indicates that 75% or fewer exam takers get the question correct)

9. Splices and taps are permitted within a nonmetallic wireway provided they are accessible. The conductors, including splices and taps, must not fill the wireway to more than _____ percent of its area at that point.

   (a) 25 (b) 80 (c) 125 (d) 75

10. Where equipment grounding is required by Article 250 for nonmetallic wireway installations, a separate equipment grounding conductor must _____.

   (a) be run outside the raceway using solid copper wire
   (b) be installed in the raceway
   (c) be obtained using a separate driven ground rod
   (d) not be required

Article 380 Multioutlet Assemblies

Article Overview

A multioutlet assembly is a surface, flush, or freestanding raceway designed to hold conductors and receptacles, and is assembled in the field or at the factory [Article 100]. It is not limited to systems commonly referred to by trade names of "plugtrak" or "plugmold."

Questions

1. A multioutlet assembly can be installed in _____.

   (a) dry locations (b) wet locations (c) a and b (d) none of these

2. A multioutlet assembly cannot be installed _____.

   (a) in concealed locations
   (b) where subject to severe physical damage
   (c) where subject to corrosive vapors
   (d) all of these

3. Metal multioutlet assemblies can pass through a dry partition, provided no receptacle is concealed in the partition and the cover of the exposed portion of the system can be removed.

   (a) True (b) False

(• Indicates that 75% or fewer exam takers get the question correct)

Article 384 Strut-Type Channel Raceways

Article Overview

Strut-type channel raceway is a metallic raceway formed by installing a cover onto strut. Strut is usually utilized as a supporting assembly, but when paired with a matching cover can be used as a raceway.

Questions

1. A strut-type channel raceway is a metallic raceway intended to be mounted to the surface of, or suspended from, a structure with associated accessories for the installation of electrical conductors.

    (a) True   (b) False

2. A strut-type channel raceway can be installed _____.

    (a) where exposed
    (b) as a power pole
    (c) unbroken through walls, partitions, and floors
    (d) all of these

3. A strut-type channel raceway cannot be installed _____.

    (a) in concealed locations
    (b) where subject to corrosive vapors if protected solely by enamel
    (c) a or b
    (d) none of these

4. The ampacity adjustment factors of 310.15(B)(2)(a) do not apply to conductors installed in strut-type channel raceways where _____.

    (a) the cross-sectional area of the raceway is at least 4 sq in.
    (b) the number of current-carrying conductors do not exceed 30
    (c) the sum of the cross-sectional areas of all contained conductors does not exceed 20 percent of the interior cross-sectional area of the strut-type channel raceways
    (d) all of these

5. A surface mount strut-type channel raceway must be secured to the mounting surface with retention straps external to the channel at intervals not exceeding _____ and within 3 ft of each outlet box, cabinet, junction box, or other channel raceway termination.

    (a) 3 ft   (b) 5 ft   (c) 6 ft   (d) 10 ft

6. Splices and taps are permitted within a strut-type channel raceway provided they are accessible. The conductors, including splices and taps, must not fill the raceway to more than _____ percent of its area at that point.

    (a) 25   (b) 80   (c) 125   (d) 75

7. Strut-type channel raceway enclosures must have a means for connecting an equipment grounding conductor. The raceway is permitted as an equipment grounding conductor in accordance with 250.118(14).

    (a) True   (b) False

(• Indicates that 75% or fewer exam takers get the question correct)

Article 386 Surface Metal Raceways

Article Overview

A surface metal raceway is a common method of adding a raceway when exposed conduit systems are not acceptable and concealing the raceway is not economically feasible. It comes in several colors, and is now available with colored or real wood inserts designed to make it look like molding rather than a raceway.

Questions

1. It is permissible to run unbroken lengths of surface metal raceways through dry _____.

    (a) walls   (b) partitions   (c) floors   (d) all of these

2. Surface metal raceways must not be used _____.

    (a) where subject to severe physical damage
    (b) where subject to corrosive vapors
    (c) in hoistways
    (d) all of these

3. The adjustment factors of 310.15(B)(2)(a), (Notes to Ampacity Tables of 0 through 2,000V), do not apply to conductors installed in surface metal raceways where _____.

    (a) the cross-sectional area exceeds 4 sq in.
    (b) the current-carrying conductors do not exceed 30 in number
    (c) the total cross-sectional area of all conductors does not exceed 20 percent of the interior cross-sectional area of the raceway
    (d) all of these

4. •The maximum number of conductors permitted in any surface raceway must be _____.

    (a) no more than 30 percent of the inside diameter
    (b) no greater than the number for which it was designed
    (c) no more than 75 percent of the cross-sectional area
    (d) that which is permitted in the Table 312.6(A)

5. Surface metal raceways must be secured and supported at intervals _____.

    (a) in accordance with the manufacturer's installation instructions
    (b) appropriate for the building design
    (c) not exceeding 8 ft
    (d) not exceeding 4 ft

6. The conductors, including splices and taps, in a metal surface raceway having a removable cover must not fill the raceway to more than _____ percent of its cross-sectional area at that point.

    (a) 75   (b) 40   (c) 38   (d) 53

7. Surface metal raceway enclosures providing a transition from other wiring methods must have a means for connecting a(n) _____.

    (a) grounded neutral conductor
    (b) ungrounded conductor
    (c) equipment grounding conductor
    (d) all of these

(• Indicates that 75% or fewer exam takers get the question correct)

8. Where combination surface metal raceways are used for both signaling and for lighting and power circuits, the different systems must be run in separate compartments identified by _____ of the interior finish.

   (a) stamping (b) imprinting (c) color-coding (d) any of these

9. Surface metal raceways and their fittings must be so designed that the sections can be _____.

   (a) electrically coupled together
   (b) mechanically coupled together
   (c) installed without subjecting the wires to abrasion
   (d) all of these

(• Indicates that 75% or fewer exam takers get the question correct)

Article 388 Surface Nonmetallic Raceways

Article Overview

A surface nonmetallic raceway is a common method of adding a raceway when exposed conduit systems are not acceptable and concealing the raceway is not economically feasible. Surface nonmetallic raceway is less expensive than a comparable surface metallic raceway and more easily installed, but may not be as impact resistant. However, it does not dent, deform, or lose paint like the metallic versions, so it may retain its appearance longer.

Questions

1. It is permissible to run unbroken lengths of surface nonmetallic raceways through dry _____.

   (a) walls   (b) partitions   (c) floors   (d) all of these

2. The maximum number of conductors permitted in any surface nonmetallic raceway must be _____.

   (a) no more than 30 percent of the inside diameter   (b) no greater than the number for which it was designed
   (c) no more than 75 percent of the cross-sectional area   (d) that which is permitted in Table 312.6(A)

3. The conductors, including splices and taps, in a surface nonmetallic raceway having a removable cover, must not fill the raceway to more than _____ percent of its cross-sectional area at that point.

   (a) 75   (b) 40   (c) 38   (d) 53

4. Where combination surface nonmetallic raceways are used for both signaling conductors and for lighting and power circuits, the different systems must be run in separate compartments identified by _____ of the interior finish.

   (a) stamping   (b) imprinting   (c) color coding   (d) any of these

(• Indicates that 75% or fewer exam takers get the question correct)

Article 392 Cable Trays

Article Overview

A cable tray system is a unit or assembly of units or sections with associated fittings that forms a structural system used to securely fasten or support cables and raceways. Cable systems include ladder, ventilated trough, ventilated channel, solid bottom, and other similar structures.

Questions

1. A cable tray is a unit or assembly of units or sections and associated fittings forming a _____ system used to securely fasten or support cables and raceways.

   (a) structural   (b) flexible   (c) movable   (d) secure

2. Cable trays can be used as a support system for _____.

   (a) services, feeders, and branch circuits   (b) communications circuits
   (c) control and signaling circuits   (d) all of these

3. •The intent of Article 392 is to limit the use of cable trays to industrial establishments only.

   (a) True   (b) False

4. Where exposed to direct rays of the sun, insulated conductors and jacketed cables must be _____ as being sunlight resistant.

   (a) listed   (b) approved   (c) identified   (d) none of these

5. Cable trays and their associated fittings must be _____ for the intended use.

   (a) listed   (b) approved   (c) identified   (d) none of these

6. Nonmetallic cable trays are permitted in corrosive areas and in areas requiring voltage isolation.

   (a) True   (b) False

7. Cable tray systems must not be used _____.

   (a) in hoistways   (b) where subject to severe physical damage
   (c) in hazardous locations   (d) a and b

8. Each run of cable tray must be _____ before the installation of cables.

   (a) tested for 25 ohms resistance   (b) insulated
   (c) completed   (d) all of these

9. Supports for cable trays must be provided in accordance with _____.

   (a) installation instructions   (b) the *NEC*   (c) a or b   (d) none of these

(• Indicates that 75% or fewer exam takers get the question correct)

10. Cable trays can extend through partitions and walls or vertically through platforms and floors where the installation is made in accordance with the fire seal requirements of 300.21.

    (a) True   (b) False

11. Cable trays must be _____ except as permitted by 392.6(G).

    (a) exposed   (b) accessible   (c) concealed   (d) a and b

12. In industrial facilities where conditions of maintenance and supervision ensure that only qualified persons will service the installation, cable tray systems can be used to support _____.

    (a) raceways   (b) cables   (c) boxes and conduit bodies   (d) all of these

13. For raceways terminating at the tray, a(n) _____ cable tray clamp or adapter must be used to securely fasten the raceway to the cable tray system.

    (a) listed   (b) approved   (c) identified   (d) none of these

14. Steel or aluminum cable tray systems can used as an equipment grounding conductor provided the cable tray sections and fittings are identified for _____ purposes, among other requirements.

    (a) grounding   (b) special   (c) industrial   (d) all

15. Aluminum cable trays must not be used as an equipment grounding conductor for circuits with ground-fault protection above _____.

    (a) 2,000A   (b) 300A   (c) 500A   (d) 1,200A

16. Steel cable trays must not be used as equipment grounding conductors for circuits with ground-fault protection above _____.

    (a) 200A   (b) 300A   (c) 600A   (d) 800A

17. One of the requirements that must be met to use steel or aluminum cable tray systems as equipment grounding conductors, is that the cable tray sections and fittings have been _____ marked to show the cross-sectional area of metal in channel cable trays, or cable trays of one-piece construction and total cross sectional area of both side rails for ladder or trough cable trays.

    (a) legibly   (b) durably   (c) a or b   (d) a and b

(• Indicates that 75% or fewer exam takers get the question correct)

Questions for Chapter 4—Equipment for General Use

Article 400 Flexible Cords and Flexible Cables

Article Overview

This article covers the general requirements, applications, and construction specifications for flexible cords and flexible cables. The *NEC* doesn't consider flexible cords to be wiring methods like those defined in Chapter 3.

Always use a cord (and fittings) identified for the application. For example, use cords listed for a wet location if you're using the cord outdoors in a wet location. The jacket material of any cord is tested to maintain its insulation properties and other characteristics only in the environments for which it has been listed.

The accompanying textbook doesn't go into detail about Table 400.4, but you should take a few moments to review it. It's not limited to extension cords—three of the entries are for elevator cables. You do not need to memorize this table, but do become aware of the types of cords and cables it covers.

Questions

1. The allowable ampacity of flexible cords and cables is found in _____.

    (a) Table 310.16   (b) Table 400.5(A) and (B)   (c) Table 1, Chapter 9   (d) Table 430.52

2. Where flexible cords are used in ambient temperatures exceeding _____ the temperature correction factors from Table 310.16 that correspond to the temperature rating of the cord must be applied to the ampacity from Table 400.5(A) or 400.5(B).

    (a) 60°C   (b) 30°C   (c) 75°C   (d) 90°C

3. A 3-conductor 16 AWG, SJE cable (one conductor is used for grounding) has a maximum ampacity of _____ for each conductor.

    (a) 13A   (b) 12A   (c) 15A   (d) 8A

4. Flexible cords and cables can be used for _____.

    (a) wiring of luminaires
    (b) connection of portable lamps or appliances
    (c) connection of utilization equipment to facilitate frequent interchange
    (d) all of these

(• Indicates that 75% or fewer exam takers get the question correct)

5. Flexible cords must not be used as a substitute for _____ wiring unless specifically permitted in 400.7.

   (a) temporary (b) fixed (c) overhead (d) none of these

6. Unless specifically permitted in 400.7, flexible cords and cables must not be used where _____.

   (a) run through holes in walls, ceilings, or floors
   (b) run through doorways, windows, or similar openings
   (c) attached to building surfaces
   (d) all of these

7. Flexible cords and cables must not be concealed behind building _____, or run through doorways, windows, or similar openings.

   (a) structural ceilings
   (b) suspended or dropped ceilings
   (c) floors or walls
   (d) all of these

8. Flexible cords and cables must be connected to devices and to fittings so that tension will not be transmitted to joints or terminal screws. This must be accomplished by _____.

   (a) knotting the cord
   (b) winding the cord with tape
   (c) fittings designed for the purpose
   (d) any of these

9. In industrial establishments where conditions of maintenance and supervision ensure that only qualified persons service the installation, flexible cords and cables are permitted to be installed in aboveground raceways that are no longer than _____, to protect the flexible cord or cable from physical damage.

   (a) 25 ft (b) 50 ft (c) 100 ft (d) no limit

(• Indicates that 75% or fewer exam takers get the question correct)

Article 402 Fixture Wires

Article Overview

This article covers the general requirements and construction specifications for fixture wires. One such requirement is that no fixture wire can be smaller than 18 AWG. Another requirement is that fixture wires must be of a type listed in Table 402.3. That table makes up the bulk of Article 402.

Questions

1. 18 TFFN has an ampacity of _____.

    (a) 14A (b) 10A (c) 8A (d) 6A

2. The smallest size fixture wire permitted in the *NEC* is _____ AWG.

    (a) 22 (b) 20 (c) 18 (d) 16

3. The number of fixture wires in a single conduit or tubing must not exceed that permitted by the percentage fill specified in _____.

    (a) Table 1, Chapter 9 (b) Table 250.66 (c) Table 310.16 (d) 240.6

4. Fixture wires are used to connect luminaires to the _____ conductors supplying the luminaires.

    (a) service (b) branch-circuit (c) feeder (d) none of these

5. Fixture wires are permitted for installation in luminaires and in similar equipment where enclosed or protected and not subject to _____ in use, or for connecting luminaires to the branch-circuit conductors supplying the luminaires.

    (a) bending or twisting (b) knotting (c) stretching or straining (d) none of these

6. Fixture wires cannot be used for branch-circuit wiring.

    (a) True (b) False

7. _____ protection for fixture wires must be as specified in 240.5

    (a) Arc-fault (b) Overcurrent (c) Ground-fault (d) Lightning

(• Indicates that 75% or fewer exam takers get the question correct)

Article 404 Switches

Article Overview

The requirements of Article 404 apply to switches of all types, such as snap (toggle) switches, dimmers, fan switches, knife switches, circuit breakers used as switches, and automatic switches. Automatic switches include those used as time clocks and timers, plus switches and circuit breakers used for disconnecting means. Here are a few key points to remember:

- Enclosures for switches or circuit breakers can contain splices if you meet certain conditions.
- Observe the wet location requirements. These include locations subject to saturation with water such as those near some showers, tubs, and pools.
- Observe switch grouping and accessibility requirements.
- Observe requirements for mounting, marking, grounding (bonding), orientation, rating, and labeling of various kinds of switches.

Questions

1. Three-way and four-way switches must be wired so that all switching is done only in the _____ circuit conductor.

   (a) ungrounded   (b) grounded   (c) equipment ground   (d) neutral

2. When grouping conductors of three-way and four-way switch loops in the same raceway to avoid inductive heating according to 300.20(A), it is not necessary to include a grounded neutral conductor in every switch loop.

   (a) True   (b) False

3. Switches or circuit breakers must not disconnect the grounded neutral conductor of a circuit unless the switch or circuit breaker _____.

   (a) can be opened and closed by hand levers only
   (b) simultaneously disconnects all conductors of the circuit
   (c) opens the grounded neutral conductor before it disconnects the ungrounded conductors
   (d) none of these

4. Switch or circuit-breaker enclosures can be used as a junction box or raceway for conductors feeding through splices or taps, when installed in accordance with 312.8.

   (a) True   (b) False

5. Switches or circuit breakers in a wet location or outside of a building must be enclosed in a _____ enclosure or cabinet that complies with 312.2(A).

   (a) weatherproof   (b) rainproof   (c) watertight   (d) raintight

6. Switches must not be installed within wet locations in tub or shower spaces unless installed as part of a listed tub or shower assembly.

   (a) True   (b) False

(• Indicates that 75% or fewer exam takers get the question correct)

Mike Holt Enterprises, Inc. • www.NECcode.com • 1.888.NEC.Code

7. Single-throw knife switches must be installed so that gravity will tend to close the switch.

   (a) True   (b) False

8. Which of the following switches must indicate whether they are in the open (off) or closed (on) position?

   (a) General-use switches.   (b) Motor-circuit switches.   (c) Circuit breakers.   (d) all of these

9. All switches and circuit breakers used as switches must be installed so that they may be operated from a readily accessible place. They must be installed so that the center of the grip of the operating handle of the switch or circuit breaker, when in its highest position, is not more than 6 ft 7 in. above the floor or working platform.

   (a) True   (b) False

10. Switches and circuit breakers used as switches can be mounted _____ if they are installed adjacent to motors, appliances, or other equipment that they supply and are accessible by portable means.

    (a) never more than 6 ft 7 in.
    (b) higher than the standard maximum of 6 ft 7 in.
    (c) only in the mechanical equipment room
    (d) up to 8 ft high

11. Snap switches must not be grouped or ganged in enclosures with other _____ if the voltage between adjacent devices exceeds 300V, unless identified barriers are securely installed between adjacent devices.

    (a) snap switches   (b) receptacles   (c) similar devices   (d) all of these

12. Snap switches must not be grouped or ganged in enclosures unless they can be arranged so that the voltage between adjacent devices does not exceed _____, or unless they are installed in enclosures equipped with permanently installed barriers between adjacent devices.

    (a) 100V   (b) 200V   (c) 300V   (d) 400V

13. All snap switches, including dimmer and similar control switches, must be effectively grounded so that they can provide a means to ground metal faceplates, whether or not a metal faceplate is installed.

    (a) True   (b) False

14. A snap switch without a grounding connection is allowed for replacement purposes only where the wiring method does not include an equipment ground and must be _____.

    (a) provided with a faceplate of nonconducting, noncombustible material
    (b) protected by a ground-fault circuit interrupter
    (c) a or b
    (d) none of these

15. Snap switches installed in recessed boxes must have the _____ seated against the finished wall surface.

    (a) mounting yoke   (b) body   (c) toggle   (d) all of these

16. A hand-operable circuit breaker equipped with a _____, or a power operated circuit breaker capable of being opened by hand in the event of a power failure, is permitted to serve as a switch if it has the required number of poles.

    (a) lever   (b) handle   (c) shunt trip   (d) a or b

17. Metal enclosures for switches or circuit breakers must be _____ as specified in Article 250.

    (a) ventilated   (b) dust proof   (c) grounded   (d) sealed

(• Indicates that 75% or fewer exam takers get the question correct)

18. Nonmetallic enclosures for switches and circuit breakers must be installed with a wiring method that provides or includes _____.

    (a) a grounded neutral conductor
    (b) an equipment ground
    (c) an inductive balance
    (d) none of these

19. Alternating current general-use snap switches, suitable only for use on ac circuits, can control _____.

    (a) resistive and inductive loads that do not exceed the ampere and voltage rating of the switch
    (b) tungsten-filament lamp loads that do not exceed the ampere rating of the switch at 120V
    (c) motor loads that do not exceed 80 percent of the ampere and voltage rating of the switch
    (d) all of these

20. AC or DC general-use snap switches, suitable for use on either ac or dc circuits, may be used for control of inductive loads not exceeding _____ percent of the ampere rating of the switch at the applied voltage.

    (a) 75
    (b) 90
    (c) 100
    (d) 50

21. Snap switches rated _____ or less directly connected to aluminum conductors must be listed and marked CO/ALR.

    (a) 15A
    (b) 20A
    (c) 25A
    (d) 30A

22. Switches must be marked with _____.

    (a) current
    (b) voltage
    (c) maximum horsepower, if horsepower rated
    (d) all of these

23. A switching device with a marked "OFF" position must completely disconnect all _____ conductors of the load it controls.

    (a) grounded
    (b) ungrounded
    (c) grounding
    (d) all of these

(• Indicates that 75% or fewer exam takers get the question correct)

Article 406 Receptacles, Cord Connectors, and Attachment Plugs (Caps)

Article Overview

This article covers the rating, type, and installation of receptacles, cord connectors, and attachment plugs (cord caps). It also addresses their grounding (bonding) requirements. Some key points to remember include:

- Follow the grounding (bonding) requirements of the specific type of device you're using.
- Use GFCIs where specified by 406.3(D)(2), and install them according to manufacturer's instructions.
- Mount receptacles according to the requirements of 406.4. These are highly detailed.

Questions

1. Receptacles and cord connectors must be rated not less than _____ at 125 volts or at 250 volts, and must be of a type not suitable for use as lampholders.

   (a) 30A (b) 20A (c) 15A (d) 10A

2. Receptacles rated 20A or less and designed for the direct connection of aluminum conductors must be marked _____.

   (a) aluminum rated (b) alum 20a (c) CO/ALR (d) al wire

3. Receptacles incorporating an isolated grounding connection intended for the reduction of electrical noise must be identified by _____ on the face of the receptacle.

   (a) an orange triangle
   (c) a completely orange device
   (b) a green triangle
   (d) the engraved word "ISOLATED"

4. Isolated ground receptacles installed in nonmetallic boxes must be covered with a nonmetallic faceplate because a metal faceplate cannot be connected to the circuit equipment grounding conductor.

   (a) True (b) False

5. Receptacles and cord connectors having grounding terminals must have those terminals effectively _____.

   (a) grounded (b) bonded (c) labeled (d) listed

6. When replacing a receptacle, and the grounding means exists in the receptacle enclosure, or a grounding conductor is installed in accordance with 250.130(C), _____-type receptacles must be used.

   (a) isolated ground (b) grounding (c) GFCI (d) two-wire

7. When replacing receptacles in locations that would require GFCI protection under the current *Code*, _____ receptacles must be installed.

   (a) two wire (b) isolated ground (c) GFCI-protected (d) grounding

8. •When replacing a nongrounding type receptacle in a bedroom of a dwelling unit where no grounding means exists in the receptacle enclosure, you must use a _____.

   (a) nongrounding receptacle (b) grounding receptacle (c) GFCI-type receptacle (d) a or c

(• Indicates that 75% or fewer exam takers get the question correct)

9. Receptacles connected to circuits having different voltages, frequencies, or types of current (ac or dc) on the _____ must be of such design that the attachment plugs used on these circuits are not interchangeable.

   (a) building   (b) interior   (c) same premises   (d) exterior

10. Receptacles mounted in boxes set back of the wall surface must be installed so that the mounting _____ of the receptacle is/are held rigidly at the surface of the wall.

    (a) screws or nails   (b) yoke or strap   (c) face plate   (d) none of these

11. Receptacles mounted in boxes flush with the wall surface or projecting beyond it must be installed so that the mounting yoke or strap of the receptacle is _____.

    (a) held rigidly against the box or box cover
    (b) mounted behind the wall surface
    (c) held rigidly at the finished surface
    (d) none of these

12. Receptacles mounted to and supported by a cover must be secured by more than one screw unless listed and identified for securing by a single screw.

    (a) True   (b) False

13. Receptacles in countertops and similar work surfaces in dwelling units must not be installed _____.

    (a) in the sides of cabinets
    (b) in a face-up position
    (c) on GFCI circuits
    (d) on the kitchen small-appliance circuit

14. Metal faceplates for receptacles must be grounded.

    (a) True   (b) False

15. Attachment plugs and cord connectors must be listed for the purpose and marked with the _____.

    (a) manufacturer's name or identification
    (b) voltage rating
    (c) amperage rating
    (d) all of these

16. Attachment plugs, cord connectors, and flanged-surface devices, must be listed with the manufacturer's name or identification and voltage and ampere ratings.

    (a) True   (b) False

17. Attachment plugs must be installed so that their prongs, blades, or pins are not energized unless inserted into an energized receptacle. No receptacle can be installed so as to require an energized attachment plug as its _____.

    (a) load   (b) source of supply   (c) protective device   (d) none of these

18. Receptacles, cord connectors, and attachment plugs must be constructed so that the receptacles or cord connectors do not accept an attachment plug with a different _____ or current rating than that for which the device is intended.

    (a) voltage rating
    (b) amperage interrupting capacity
    (c) temperature rating
    (d) all of these

19. A receptacle is considered to be in a location protected from the weather when located under roofed open porches, canopies, marquees, and the like, where it will not be subjected to _____.

    (a) spray from a hose
    (b) a direct lightning hit
    (c) beating rain or water runoff
    (d) falling or wind-blown debris

(• Indicates that 75% or fewer exam takers get the question correct)

20. Receptacles installed outdoors, in a location protected from the weather or other damp locations, must be in an enclosure that is _____ when the receptacle is covered.

    (a) raintight   (b) weatherproof   (c) rainproof   (d) weathertight

21. An outdoor receptacle in a location protected from the weather, or another damp location, must be installed in an enclosure that is weatherproof when the receptacle is _____.

    (a) covered   (b) enclosed   (c) protected   (d) none of these

22. A receptacle is considered to be in a location protected from the weather (damp location) where _____.

    (a) located under a roofed open porch
    (b) not subjected to beating rain or water runoff
    (c) a or b
    (d) a and b

23. _____, 125 and 250V receptacles installed in a wet location must have an enclosure that is weatherproof whether or not the attachment plug cap is inserted.

    (a) 15A   (b) 20A   (c) a and b   (d) none of these

24. Which of the following statements are true for covers in a wet location for 15 and 20A, 125 and 250V receptacles?
    (1) The receptacle cover for a product, which will not be attended while in use, must be listed as weatherproof while the attachment plug is inserted or removed.
    (2) Portable loads such as appliances and power tools must make use of a receptacle cover that is listed as weatherproof while the attachment plug is removed.

    (a) 1 only   (b) 2 only   (c) 1 and 2   (d) none of these

25. An enclosure that is weatherproof, only when no attachment plug is connected, can be used for receptacles in a wet location other than outdoors when the receptacle is used for _____ while attended.

    (a) portable equipment   (b) portable tools   (c) fixed equipment   (d) a and b

26. A receptacle must not be installed within, or directly over, a bathtub or shower space.

    (a) True   (b) False

27. A receptacle installed in an outlet box flush-mounted on a finished surface in a damp or wet location must be made weatherproof by means of a weatherproof faceplate assembly that provides a _____ connection between the plate and the finished surface.

    (a) sealed   (b) weathertight   (c) sealed and protected   (d) watertight

28. Grounding-type attachment plugs must be used only with a cord having a(n) _____ conductor.

    (a) equipment grounding   (b) isolated   (c) computer circuit   (d) insulated

(• Indicates that 75% or fewer exam takers get the question correct)

Article 408 Switchboards and Panelboards

Article Overview

Article 408 covers the specific requirements for switchboards, panelboards, and distribution boards that control light and power circuits. Some key points to remember:

- One objective of Article 408 is that the installation prevents contact between current-carrying conductors and people or maintenance equipment.
- The circuit directory of a panelboard must clearly identify the purpose or use of each circuit that originates in that panelboard.
- You must know the difference between a "lighting and appliance panelboard" and a "power panelboard."
- You must understand the detailed grounding (bonding) requirements for panelboards.

Questions

1. Panelboards supplied by a 3Ø, 4-wire, delta-connected system must have that phase with the higher voltage-to-ground (high-leg) connected to the _____ phase.

   (a) A   (b) B   (c) C   (d) any of these

2. The purpose or use of panelboard circuits and circuit _____ must be legibly identified on a circuit directory located on the face or inside of the doors of a panelboard, and at each switch on a switchboard.

   (a) manufacturers   (b) conductors   (c) feeders   (d) modifications

3. Conduits and raceways, including end fittings, must not rise more than _____ above the bottom of a switchboard enclosure.

   (a) 3 in.   (b) 4 in.   (c) 5 in.   (d) 6 in.

4. Unused openings for circuit breakers and switches in switchboards and panelboards must be closed using _____ or other approved means that provide protection substantially equivalent to the wall of the enclosure.

   (a) duct seal and tape   (b) identified closures   (c) exothermic welding   (d) sheet metal

5. To qualify as a lighting and appliance branch-circuit panelboard, the number of circuits rated at 30A or less and having a neutral conductor must be _____ of the total.

   (a) more than 10 percent   (b) 10 percent   (c) 20 percent   (d) 40 percent

6. Not counting the main breaker, the maximum number of overcurrent devices that can be installed in any one cabinet of a lighting and appliance branch-circuit panelboard is _____.

   (a) 12   (b) 42   (c) 6   (d) none of these

7. A lighting and appliance branch-circuit panelboard must be provided with physical means to prevent the installation of more _____ devices than that number for which the panelboard was designed, rated, and approved.

   (a) overcurrent   (b) equipment   (c) circuit breaker   (d) all of these

(• Indicates that 75% or fewer exam takers get the question correct)

8. A lighting and appliance branch-circuit panelboard contains six 3-pole breakers and eight 2-pole breakers. The maximum allowable number of single-pole breakers that can be added to this panelboard is _____.

   (a) 8 (b) 16 (c) 28 (d) 42

9. A lighting and appliance branch-circuit panelboard is not required to be individually protected if the panelboard _____ conductor has protection not greater than the panelboard rating.

   (a) grounded neutral (b) feeder (c) branch circuit (d) none of these

10. When a lighting and appliance branch-circuit panelboard is supplied from a transformer, the overcurrent protection must be located _____.

    (a) on the secondary side of the transformer
    (c) none is required
    (b) on the primary side of the transformer
    (d) either a or b

11. Plug-in-type circuit breakers that are back-fed (to supply a panelboard) must be _____ by an additional fastener that requires more than a pull to release.

    (a) grounded (b) secured in place (c) shunt tripped (d) none of these

12. When equipment grounding conductors are installed in panelboards, a _____ is required for the proper termination of the equipment grounding conductors.

    (a) grounded neutral conductor
    (c) grounding terminal bar
    (b) terminator strip
    (d) none of these

13. Each _____ conductor must terminate within the panelboard at an individual terminal that is not also used for another conductor.

    (a) grounded (b) ungrounded (c) grounding (d) all of these

(• Indicates that 75% or fewer exam takers get the question correct)

Article 410 Luminaires, Lampholders, and Lamps

Article Overview

This article covers luminaires, lampholders, pendants, and the wiring and equipment of such lamps and luminaires. The scope of Article 410 has been expanded to include decorative lighting products, lighting accessories for temporary seasonal and holiday use, and portable flexible lighting products.

This article is highly detailed, but it's broken down into sixteen parts. The first five are sequential, and apply to all luminaires, lampholders, and lamps—General, Location, Boxes and Covers, Supports, and Grounding (Bonding). This is mostly mechanical information, and it's not hard to follow or absorb. Part VI, Wiring, ends the sequence. The seventh, ninth, and tenth parts provide requirements for manufacturers to follow—use only equipment that conforms to these requirements. Part VIII provides requirements for Installing Lampholders. The rest of Article 410 addresses specific types of lighting.

Author's Comment: Article 410 doesn't include "Lighting Systems Operating at 30 Volts or Less"; Article 411 addresses them.

Questions

1. Article 410 covers luminaires, lampholders, pendants, and _____, and the wiring and equipment forming part of such products and lighting installations.

   (a) decorative lighting products
   (b) lighting accessories for temporary seasonal and holiday use
   (c) portable flexible lighting products
   (d) all of these

2. A luminaire marked "Suitable for Damp Locations" _____ be used in a wet location.

   (a) can   (b) cannot

3. Luminaires are permitted to be installed in a commercial cooking hood where specific conditions are met, including the requirement that the luminaire be identified for use within a _____ cooking hood.

   (a) nonresidential   (b) commercial   (c) multifamily   (d) all of these

4. No part of cord-connected luminaires, hanging luminaires, track lighting, pendants, or paddle fans may be located within a zone measured 3 ft horizontally and _____ vertically from the top of the bathtub rim or shower stall threshold.

   (a) 4 ft   (b) 6 ft   (c) 8 ft   (d) none of these

5. Luminaires located in bathtub and shower zones must be listed for damp locations, or listed for wet locations where _____.

   (a) below 7 ft in height   (b) below 6 ft 7 in. in height   (c) subject to shower spray   (d) not GFCI protected

6. Luminaires using a _____ lamp, that are subject to physical damage and installed in playing and spectator seating areas of indoor sports, mixed-use, or all-purpose facilities, must be of the type that protects the lamp with a glass or plastic lens. Such luminaires are permitted to have an additional guard.

   (a) mercury vapor   (b) metal halide   (c) fluorescent   (d) a or b

(• Indicates that 75% or fewer exam takers get the question correct)

7. The *NEC* requires a lighting outlet in clothes closets.

   (a) True   (b) False

8. Incandescent luminaires that have open lamps, and pendant-type luminaires, can be installed in clothes closets where proper clearance is maintained from combustible products.

   (a) True   (b) False

9. Surface-mounted fluorescent luminaires in clothes closets can be installed on the wall above the door, or on the ceiling, provided there is a minimum clearance of _____ between the luminaire and the nearest point of a storage space.

   (a) 3 in.   (b) 6 in.   (c) 9 in.   (d) 12 in.

10. In clothes closets, recessed incandescent luminaires with a completely enclosed lamp are permitted to be installed in the wall or on the ceiling, provided there is a minimum clearance of _____ between the luminaire and the nearest point of a storage space.

    (a) 3 in.   (b) 6 in.   (c) 9 in.   (d) 12 in.

11. Electric-discharge luminaires supported independently of the outlet box must be connected to the branch circuit through _____.

    (a) raceways
    (c) flexible cords
    (b) Type MC, AC, MI, NM, or NMC cable
    (d) a, b, or c

12. When an electric-discharge luminaire is mounted directly over a concealed outlet box, which is not its sole means of support, the luminaire must provide access to the conductor wiring within the outlet box by means of suitable openings in the back of the fixture.

    (a) True   (b) False

13. The maximum weight of a luminaire that can be supported by the screw-shell of a lampholder is _____.

    (a) 2 lbs   (b) 6 lbs   (c) 3 lbs   (d) 50 lbs

14. Handholes in metal or nonmetallic poles supporting luminaires are not required for poles _____ or less in height above finished grade, if the pole is provided with a hinged base and the grounding terminal is accessible within the hinged base.

    (a) 8 ft   (b) 18 ft   (c) 20 ft   (d) none of these

15. Metal or nonmetallic poles over 20 ft in height above grade that support luminaires must meet which of the following requirements? _____.

    (a) They must have an accessible handhole (sized 2 x 4 in.) with a raintight cover
    (b) The grounding terminal must be accessible from the handhole
    (c) a and b
    (d) none of these

16. Metal poles used to support luminaires must be bonded to a(n) _____.

    (a) grounding electrode
    (c) equipment grounding conductor
    (b) grounded neutral conductor
    (d) any of these

(• Indicates that 75% or fewer exam takers get the question correct)

17. Luminaires attached to the framing of a suspended-ceiling must be secured to the framing member(s) by mechanical means such as bolts, screws, or rivets. Clips _____ and identified for use with the type of ceiling framing member(s) and luminaires are also permitted.

    (a) marked   (b) labeled   (c) identified   (d) listed

18. Trees can to be used to support outdoor luminaires.

    (a) True   (b) False

19. Luminaires that require adjustment or aiming after installation can be cord-connected without an attachment plug.

    (a) True   (b) False

20. A listed luminaire or a listed assembly is permitted to be cord-connected if located _____ the outlet box, the cord is continuously visible for its entire length outside the luminaire, and the cord is not subject to strain or physical damage.

    (a) within   (b) directly below   (c) directly above   (d) adjacent to

21. The flexible cord used to connect a luminaire may be terminated _____.

    (a) in a grounding-type attachment plug cap
    (b) as a part of a listed assembly incorporating a manufactured wiring system connector
    (c) as a part of a listed luminaire assembly with a strain relief and canopy
    (d) all of these

22. Luminaires designed for end-to-end connection to form a continuous assembly, or luminaires connected together by recognized wiring methods, are permitted to contain the conductors of a 2-wire branch circuit, or one _____ branch circuit, supplying the connected luminaires and need not be listed as a raceway.

    (a) small-appliance   (b) appliance   (c) multiwire   (d) industrial

23. Branch-circuit conductors within _____ of a ballast must have an insulation temperature rating not lower than 90°C (194°F) unless supplying a luminaire that is listed and marked as suitable for a different insulation temperature.

    (a) 1 in.   (b) 3 in.   (c) 6 in.   (d) none of these

24. Edison-base screw-shell lampholders are designed to hold lamps and to hold screw-in receptacle adapters.

    (a) True   (b) False

25. A recessed incandescent luminaire (fixture) must be installed so that adjacent combustible material will not be subjected to temperatures in excess of _____°C.

    (a) 75   (b) 90   (c) 125   (d) 150

26. Recessed incandescent luminaires must have _____ protection and must be identified as thermally protected.

    (a) physical   (b) corrosion   (c) thermal   (d) all of these

27. A recessed luminaire (fixture) that is not identified for contact with insulation must have all recessed parts spaced not less than _____ from combustible materials, except for points of support and the trim finishing off the opening in the ceiling or wall.

    (a) ¼ in.   (b) ½ in.   (c) 1¼ in.   (d) 6 in.

(• Indicates that 75% or fewer exam takers get the question correct)

28. A Type IC recessed luminaire, which is identified for contact with insulation, is permitted to be in contact with _____.

   (a) combustible material at recessed parts
   (b) points of support
   (c) portions passing through or finishing off the opening in the building structure
   (d) all of these

29. Thermal insulation must not be installed above a recessed luminaire or within _____ of the recessed luminaire's enclosure, wiring compartment, or ballast unless it is a Type IC luminaire.

   (a) 6 in. (b) 12 in. (c) 3 in. (d) ½ in.

30. The minimum distance that an outlet box containing tap supply conductors can be placed from a recessed luminaire (fixture) is _____.

   (a) 1 ft (b) 2 ft (c) 3 ft (d) 4 ft

31. The raceway or cable for tap conductors to recessed luminaires must have a minimum length of _____

   (a) 6 in. (b) 12 in. (c) 18 in. (d) 24 in.

32. Luminaires containing a metal halide lamp, other than a thick-glass parabolic reflector lamp (PAR), must be provided with a containment barrier that encloses the lamp, or the luminaire must be provided with a physical means that only allows the use of a(n) _____.

   (a) Type "O" lamp (b) Type PAR lamp (c) a or b (d) inert gas

33. In indoor locations, other than dwellings and associated accessory structures, fluorescent luminaires that utilize double-ended lamps and contain ballast(s) that can be serviced or re-ballasted in place must have a disconnecting means, to disconnect simultaneously all conductors of the ballast, including the _____ conductor if any. The disconnecting means must be accessible to qualified persons. This requirement will become effective January 1, 2008

   (a) high leg (b) grounded neutral (c) equipment ground (d) b and c

34. Surface-mounted luminaires with a ballast must have a minimum clearance of _____ from combustible low-density cellulose fiberboard, unless the fixture is marked "Suitable for Surface Mounting on Combustible Low-Density Cellulose Fiberboard."

   (a) ½ in. (b) 1 in. (c) 1½ in. (d) 2 in.

35. Electric-discharge luminaires having an open-circuit voltage exceeding _____ must not be installed in or on dwelling occupancies.

   (a) 120V (b) 250V (c) 600V (d) 1,000V

36. Lighting track is a manufactured assembly and its length may not be altered by the addition or subtraction of sections of track.

   (a) True (b) False

37. Track lighting fittings are permitted to be equipped with general-purpose receptacles.

   (a) True (b) False

38. Track lighting must not be installed _____.

   (a) where subject to physical damage
   (b) in wet or damp locations
   (c) a and b
   (d) none of these

(• Indicates that 75% or fewer exam takers get the question correct)

39. Lighting track must not be installed less than _____ above the finished floor except where protected from physical damage or track operating at less than 30 volts rms, open-circuit voltage.

    (a) 4 ft   (b) 5 ft   (c) 5½ ft   (d) 6 ft

40. Lighting track must not be installed within the zone measured 3 ft horizontally and _____ vertically from the top of the bathtub rim.

    (a) 2 ft   (b) 3 ft   (c) 4 ft   (d) 8 ft

41. Lighting track must be securely mounted so each fastening will be suitable to support the maximum weight of _____.

    (a) 35 lbs
    (c) luminaires that can be installed
    (b) 50 lbs
    (d) none of these

42. Lighting track must have 2 supports for a single section of _____ or shorter in length and each individual section of not more than 4 ft attached to it must have one additional support.

    (a) 4 ft   (b) 6 ft   (c) 10 ft   (d) 2 ft

43. Decorative lighting and similar accessories used for holiday lighting and similar purposes in accordance with 590.3(B) must be _____.

    (a) approved   (b) listed   (c) arc-fault protected   (d) all of these

(• Indicates that 75% or fewer exam takers get the question correct)

Article 411 Lighting Systems Operating at 30V or Less

Article Overview

Article 411 provides the requirements for 30V lighting systems which are found in such applications as landscaping, jewelry stores, and museums. Don't let the half-page size of Article 411 give you the impression that 30V lighting isn't something you need to be concerned about. Its applications are widespread and becoming more so.

Many of these systems now use LEDs, and 30V halogen lamps are also fairly common. All 30V lighting systems have an ungrounded secondary circuit supplied by an isolating transformer. These systems have restrictions that affect where they can be located, and they can have a maximum supply breaker size of 20A.

Questions

1. Lighting systems operating at 30V or less need not be listed for the purpose.

    (a) True (b) False

2. Lighting systems operating at 30V or less are allowed to be concealed or extended through a building wall without regard to the wiring method used.

    (a) True (b) False

3. Lighting systems operating at 30V or less are allowed to be concealed or extended through a building wall using _____.

    (a) any of the wiring methods specified in Chapter 3
    (b) wiring supplied by a listed class 2 power source installed in accordance with 725.52
    (c) both a and b
    (d) metal raceways only

4. Lighting systems operating at 30V or less must not be installed within 10 ft of pools, spas, fountains, or similar locations except as permitted by Article 680.

    (a) True (b) False

(• Indicates that 75% or fewer exam takers get the question correct)

Article 422 Appliances

Article Overview

Article 422 covers electric appliances used in any occupancy. The meat of what you need to know is in Parts II and III. Parts IV and V are primarily for manufacturers, but you should examine appliances for conformance before installing. If the appliance has a label from a recognized labeling authority (for example, UL), it conforms.

Two concepts drive the requirements of Article 422: On the one hand, an appliance should not overload the circuit supplying it; on the other hand, an appliance should not be supplied with more current than it should reasonably draw. The first concept is why 422.10 specifies the minimum circuit protection. The second concept is why 422.11 specifies the maximum circuit protection. As you read through Article 422 requirements, try to think of how each one relates to these two concepts.

Interestingly, the *NEC* doesn't include "Fixed Electric Space-Heating Equipment" in the scope of Article 422, but instead provides Article 424 to address fixed electrical equipment used for space heating.

Questions

1. In general, branch-circuit conductors to individual appliances must not be sized _____ than required by the appliance markings or instructions.

    (a) larger (b) smaller

2. Individual circuits for nonmotor-operated appliances that are continuously loaded must have the branch-circuit rating sized no less than _____ percent of the appliance marked ampere rating.

    (a) 150 (b) 100 (c) 125 (d) 80

3. If a protective device rating is marked on an appliance, the branch-circuit overcurrent protection device rating must not be greater than _____ percent of the protective device rating marked on the appliance.

    (a) 100 (b) 50 (c) 80 (d) 115

4. The rating or setting of an overcurrent protection device for a 16.3A single nonmotor-operated appliance must not exceed _____.

    (a) 15A (b) 35A (c) 25A (d) 45A

5. Central heating equipment, other than fixed electric space-heating equipment, must be supplied by a(n) _____ branch circuit.

    (a) multiwire (b) individual
    (c) multipurpose (d) small-appliance branch circuit

6. A storage-type water heater having a capacity of _____ gallons or less is considered a continuous load.

    (a) 60 (b) 75 (c) 90 (d) 120

7. A waste disposal can be cord-and-plug connected, but the cord must not be less than 18 in. or more than _____ in length and must be protected from physical damage.

    (a) 30 in. (b) 36 in. (c) 42 in. (d) 48 in.

(• Indicates that 75% or fewer exam takers get the question correct)

8. The cord for a dishwasher and trash compactor must not be longer than _____ measured from the back of the appliance.

   (a) 2 ft   (b) 4 ft   (c) 6 ft   (d) 8 ft

9. Range hoods are permitted to be cord-and-plug connected with a flexible cord identified as suitable for use on range hoods in the manufacturer's instructions when necessary conditions are met, including: _____.

   (a) The cord must not be less than 18 in. in length
   (b) The cord must be no longer than 36 in. in length
   (c) The receptacle must be supplied by an individual branch circuit
   (d) all of these

10. Ceiling-suspended (paddle) fans must be supported independently of an outlet box or by outlet boxes or outlet box systems _____ for the application and installed in accordance with 314.27(D).

    (a) satisfactory   (b) approved   (c) strong enough   (d) identified

11. The maximum allowable hp rating of a permanently connected appliance, when the branch-circuit overcurrent protection device is used as the appliance disconnecting means, is _____ or 300 VA.

    (a) ⅛ hp   (b) ¼ hp   (c) ½ hp   (d) ¾ hp

12. For permanently connected appliances rated over _____ or ⅛ hp, the branch-circuit switch or circuit breaker is permitted to serve as the disconnecting means where the switch or circuit breaker is within sight from the appliance or is capable of being locked in the open position with a permanently installed locking provision.

    (a) 200 VA   (b) 300 VA   (c) 400 VA   (d) 500 VA

13. For cord-and-plug connected appliances, an accessible separable connector or _____ plug and receptacle is permitted to serve as the disconnecting means.

    (a) a labeled   (b) an accessible   (c) a metal enclosed   (d) none of these

14. For cord-and-plug connected household electric ranges, an attachment plug and receptacle connection at the rear base of the range, if it is _____ is allowed to serve as the disconnecting means.

    (a) less than 40A   (b) a flush-mounted receptacle
    (c) GFCI-protected   (d) accessible by removal of a drawer

15. Appliances that have a unit switch with a marked _____ setting that disconnects all the ungrounded conductors is permitted to serve as the disconnecting means for the appliance, where other means of disconnection are also provided in accordance with 422.34.

    (a) "on"   (b) "off"   (c) "on/off"   (d) all of these

16. Cord-and-plug connected vending machines manufactured or remanufactured on or after January 1, 2005 must include a ground-fault circuit interrupter as an integral part of the attachment plug or in the power-supply cord within 12 in. of the attachment plug. Cord-and-plug connected vending machines not incorporating integral GFCI protection must _____.

    (a) be re-manufactured   (b) be disabled
    (c) be connected to a GFCI-protected outlet   (d) be connected to an AFCI-protected circuit

(• Indicates that 75% or fewer exam takers get the question correct)

Article 424 Fixed Electric Space-Heating Equipment

Article Overview

Many people are surprised to see how many pages are in Article 424. This is a nine-part article on fixed electric space heaters. Why so much text for what seems to be a simple application? The answer is that Article 424 covers a variety of applications—heaters come in various configurations for various uses. Not all of these Parts are for the electrician in the field—the requirements in Part IV are for manufacturers.

Most electricians should focus on Part III, Part V, and Part VI. Fixed space heaters (wall-mounted, ceiling-mounted, or free-standing) are common in many utility buildings and other small structures, as well as in some larger structures. When used to heat floors, space-heating cables address the thermal layering problem typical of forced-air systems—so it's likely you will encounter them. Duct heaters are very common in large office and educational buildings. These provide a distributed heating scheme. Locating the heater in the duct-work, but close to the occupied space, eliminates the waste of transporting heated air through sheet metal routed in unheated spaces. So, it's likely you will encounter those as well.

Questions

1. Fixed electric space-heating equipment is considered a(n) _____ load.

   (a) noncontinuous (b) intermittent (c) continuous (d) none of these

2. Permanently installed electric baseboard heaters equipped with factory-installed receptacle outlets are permitted to be used as the outlets required by 210.50(B).

   (a) True (b) False

3. Means must be provided to disconnect the _____ of all fixed electric space-heating equipment from all ungrounded conductors.

   (a) heater (b) motor controller(s)
   (c) supplementary overcurrent protective device(s) (d) all of these

4. If the disconnect is not within sight of the fixed electric space heater which includes a motor rated over ⅛ hp (without supplementary overcurrent protection devices), it must be capable of being _____.

   (a) locked (b) locked in the closed position
   (c) locked in the open position (d) within sight

5. A unit switch with a marked "off" position that is part of a fixed space heater, and disconnects all ungrounded conductors, is permitted as the disconnecting means required by Article 424 for one-family dwellings.

   (a) True (b) False

6. GFCI protection for personnel must be provided for electrically-heated floors in _____ locations.

   (a) bathroom (b) hydromassage bathtub (c) kitchen (d) a and b

7. Duct heater controller equipment must have a disconnecting means installed within _____ the controller.

   (a) 25 ft of (b) sight from (c) the side of (d) none of these

(• Indicates that 75% or fewer exam takers get the question correct)

Article 430 Motors, Motor Circuits, and Controllers

Article Overview

Article 430 contains the specific rules for conductor sizing, overcurrent protection, control circuit conductors, controllers, and disconnecting means for electric motors. The installation requirements for motor control centers are covered in Part VIII, and air-conditioning and refrigerating equipment are covered in Article 440.

Article 430 is by far the longest article in the *NEC*. It's also the most complex. But then, motors are complex. They are electrical *and* mechanical devices, but what makes motor applications complex is the fact that they are also inductive loads with a high-current demand at startup that is typically six, or more, times the running current. This makes circuit protection and motor protection necessarily different. So don't confuse circuit protection with motor protection—you must calculate and apply them separately. If you remember that as you study this Article, you will find it much easier to understand and apply.

Questions

1. For general motor applications, the motor branch-circuit short-circuit and ground-fault protection device must be sized based on the _____ amperes.

    (a) motor nameplate (b) NEMA standard (c) *NEC* Table (d) Factory Mutual

2. The motor _____ current as listed in Tables 430.247 through 430.250 must be used for sizing motor circuit conductors and short-circuit, ground-fault protection devices.

    (a) nameplate (b) full-load (c) power factor (d) service factor

3. Motor controllers and terminals of control circuit devices are required to be connected with copper conductors unless identified for use with a different type of conductor.

    (a) True (b) False

4. Torque requirements for motor control circuit device terminals must be a minimum of _____ lbs-in. (unless otherwise identified) for screw-type pressure terminals used for 14 AWG and smaller copper conductors.

    (a) 7 (b) 9 (c) 10 (d) 15

5. In determining the highest-rated motor for purposes of 430.24, the highest-rated motor must be based on the rated full-load current as selected _____.

    (a) from the motor nameplate
    (b) from tables 430.247, 430.248, 430.249, and 430.250
    (c) from taking the horsepower times 746 watts
    (d) using the largest horsepower motor

6. Branch-circuit conductors supplying a single continuous-duty motor must have an ampacity not less than _____.

    (a) 125 percent of the motor's nameplate current rating
    (b) 125 percent of the motor's full-load current as determined by 430.6(A)(1)
    (c) 125 percent of the motor's full locked-rotor rating
    (d) 80 percent of the motor's full-load current rating

(• Indicates that 75% or fewer exam takers get the question correct)

7. Conductors supplying several motors must not be sized smaller than _____ of the full-load current rating of the highest rated motor plus the sum of the full-load current ratings of all other motors in the group as determined by 430.6(A), plus the ampacity for any other loads.

   (a) 80 percent (b) 100 percent (c) 125 percent (d) 150 percent

8. Feeder tap conductors supplying motor circuits, with an ampacity of at least one-third that of the feeder, must not exceed _____ in length.

   (a) 10 ft (b) 15 ft (c) 20 ft (d) 25 ft

9. Overload devices are intended to protect motors, motor control apparatus, and motor branch-circuit conductors against _____.

   (a) excessive heating due to motor overloads (b) excessive heating due to failure to start
   (c) short circuits and ground faults (d) a and b

10. Motor overload protection is not required where _____.

    (a) conductors are oversized by 125 percent (b) conductors are part of a limited-energy circuit
    (c) it might introduce additional or increased hazards (d) short-circuit protection is provided

11. An overload device used to protect continuous-duty motors (rated more than 1 hp) must be selected to trip, or be rated, at no more than _____ percent of the motor nameplate full-load current rating for motors with a marked service factor of 1.15 or greater.

    (a) 110 (b) 115 (c) 120 (d) 125

12. The minimum number of overload unit(s) required for a 3Ø ac motor is/are _____.

    (a) one (b) two (c) three (d) any of these

13. The motor branch-circuit short-circuit and ground-fault protective device must be capable of carrying the _____ current of the motor.

    (a) varying (b) starting (c) running (d) continuous

14. The maximum rating or setting of an inverse-time breaker used as the motor branch-circuit short-circuit and ground-fault protective device for a 1Ø motor is _____ percent of the full load current given in Table 430.248.

    (a) 125 (b) 175 (c) 250 (d) 300

15. Where the motor short-circuit and ground-fault protection devices determined by Table 430.52 do not correspond to the standard sizes or ratings, a higher size may be used that does not exceed the next higher standard ampere rating.

    (a) True (b) False

16. A motor can be provided with combined overcurrent protection using a single protective device to provide branch-circuit _____ where the rating of the device provides the necessary overload protection specified in 430.32.

    (a) short-circuit protection (b) ground-fault protection (c) motor-overload protection (d) all of these

17. •A feeder must have a protective device with a rating or setting _____ branch-circuit short-circuit and ground-fault protective device for any motor in the group, plus the sum of the full-load currents of the other motors of the group.

    (a) not greater than the largest rating or setting of the (b) 125 percent of the largest rating of any
    (c) equal to the largest rating of any (d) none of these

(• Indicates that 75% or fewer exam takers get the question correct)

18. A motor control conductor that is tapped from the load side of a motor branch-circuit short-circuit and ground-fault protective device is not considered to be a branch-circuit conductor, and must be protected in accordance with 430.72.

    (a) True (b) False

19. Motor control circuits must be arranged so that they will be disconnected from all sources of supply when the disconnecting means is in the open position. Where separate devices are used for the motor and control circuit, they must be located immediately adjacent to each other.

    (a) True (b) False

20. If the control circuit transformer is located in the controller enclosure, the transformer must be connected to the _____ side of the control circuit disconnect.

    (a) line (b) load (c) adjacent (d) none of these

21. The branch-circuit protective device is permitted to serve as the controller for a stationary motor rated at _____ or less that is normally left running and cannot be damaged by overload or failure to start.

    (a) ⅛ hp (b) ¼ hp (c) ⅜ hp (d) ½ hp

22. The motor controller must have horsepower ratings at the application voltage not _____ the horsepower rating of the motor.

    (a) lower than (b) higher than (c) equal to (d) none of these

23. _____ rated in amperes is permitted as a controller for all motors.

    (a) A branch-circuit inverse-time circuit breaker
    (b) A molded-case switch
    (c) both a and b
    (d) none of these

24. For stationary motors of 2 horsepower or less and 300 volts or less on ac circuits, the controller is permitted to be an ac-rated general use snap switch where the motor full-load current rating is not more than _____ percent of the rating of the switch.

    (a) 80 (b) 50 (c) 75 (d) 125

25. The controller is required to open all conductors to the motor.

    (a) True (b) False

26. Each motor must be provided with an individual controller.

    (a) True (b) False

27. Table 430.91 provides the basis for selecting enclosures for use in specific locations other than _____ .

    (a) agricultural buildings (b) hazardous locations (c) recreational vehicle parks (d) assembly occupancies

28. A disconnecting means is required to disconnect the _____ from all ungrounded supply conductors.

    (a) motor (b) motor or controller (c) controller (d) motor and controller

29. •A _____ must be located in sight from the motor location and the driven machinery location.

    (a) controller (b) protection device (c) disconnecting means (d) all of these

(• Indicates that 75% or fewer exam takers get the question correct)

30. The motor disconnecting means is not required to be in sight from the motor and the driven machinery location, provided _____.

    (a) the controller disconnecting means is capable of being individually locked in the open position
    (b) the provisions for locking are permanently installed on, or at, the switch or circuit breaker used as the controller disconnecting means
    (c) locating the motor disconnecting means within sight of the motor is impractical or introduces additional or increased hazards to people or property
    (d) all of these

31. The disconnecting means for the controller and motor must open all ungrounded supply conductors and must be designed so that no pole can be operated independently.

    (a) True   (b) False

32. The motor disconnecting means must _____ whether it is in the open (off) or closed (on) position.

    (a) plainly indicate   (b) provide current   (c) be in the upper position   (d) none of these

33. Where more than one motor disconnecting means is provided in the same motor branch circuit, only one of the disconnecting means is required to be readily accessible.

    (a) True   (b) False

34. The motor disconnecting means can be a _____.

    (a) circuit breaker   (b) motor-circuit switch rated in horsepower
    (c) molded-case switch   (d) any of these

35. If the motor disconnecting means is a motor-circuit switch, it must be rated in _____.

    (a) horsepower   (b) watts   (c) amperes   (d) locked-rotor current

36. A branch-circuit overcurrent protection device such as a plug fuse may serve as the disconnecting means for a stationary motor of ⅛ hp or less.

    (a) True   (b) False

37. The disconnecting means for a 50 hp, 460V, 3Ø induction motor (FLC 65A) must have an ampere rating of not less than _____.

    (a) 126   (b) 75   (c) 91   (d) 63

38. A horsepower-rated inverse-time circuit breaker can serve as both a motor controller and disconnecting means if _____.

    (a) it opens all ungrounded conductors
    (b) it is protected by an overcurrent device in each ungrounded conductor
    (c) it is manually operable, or both power and manually operable
    (d) all of these

(• Indicates that 75% or fewer exam takers get the question correct)

Article 440 Air-Conditioning and Refrigerating Equipment

Article Overview

This article applies to electrically driven air-conditioning and refrigeration equipment with a hermetic refrigerant motor-compressor. The rules in this article add to, or amend, the rules in Article 430 and other articles.

Why the special treatment? Several reasons. First, hermetic motors aren't general purpose. They are sized, fitted, and engineered for specific applications. You may have seen the phrase "hermetically sealed" on food containers. Something that is hermetically sealed is airtight—no gas can enter it or escape it. These motors are sealed airtight, and that affects what you can expect of the motor in terms of performance and heat dissipation.

Each equipment manufacturer has the motors for a given air-conditioning unit built to its own specifications. Cooling and other characteristics are different from those of nonhermetic motors. For each motor, the manufacturer has worked out all of the details and supplied the correct protection, conductor sizing, and other information on the nameplate. Thus, Article 440 requires you to use the nameplate circuit protection rather than what you might develop from applying Article 430.

The application itself—with the compressor motor often on the other side of an exterior building wall from the normal power sources so it can exchange heat with free air—poses additional problems, which the *NEC* addresses in Article 440.

Three key points to remember so you apply Article 440 correctly are:

1. Hermetic motors are different. Article 440 supercedes Article 430 in regard to these motors.
2. Use the nameplate information.
3. Be careful where you install your disconnects. For motors that fall under Article 440, there are no exceptions to the rules.

Questions

1. Article 440 applies to electric motor-driven air-conditioning and refrigerating equipment that has a hermetic refrigerant motor-compressor.

   (a) True   (b) False

2. The rules of _____, as applicable, apply to air-conditioning and refrigerating equipment that do not incorporate a hermetic refrigerant motor-compressor.

   (a) Article 422   (b) Article 424   (c) Article 430   (d) all of these

3. Equipment such as _____ are considered appliances, and the provisions of Article 422 are applicable in addition to Article 440.

   (a) room air conditioners
   (b) household refrigerators and freezers
   (c) drinking water coolers and beverage dispensers
   (d) all of these

4. A disconnecting means that serves a hermetic refrigerant motor-compressor must be selected on the basis of the nameplate rated-load current or branch-circuit selection current, whichever is greater. It must have an ampere rating of at least _____ percent of the nameplate rated-load current or branch-circuit selection current, whichever is greater.

   (a) 125   (b) 80   (c) 100   (d) 115

(• Indicates that 75% or fewer exam takers get the question correct)

5. For cord-connected equipment such as _____, a separable connector or an attachment plug and receptacle is permitted to serve as the disconnecting means.

   (a) room air conditioners
   (b) household refrigerators and freezers
   (c) drinking water coolers and beverage dispensers
   (d) all of these

6. The disconnecting means for air-conditioning and refrigerating equipment must be _____ from the air-conditioning or refrigerating equipment.

   (a) readily accessible   (b) within sight   (c) a or b   (d) a and b

7. Disconnecting means must be located within sight from and readily accessible from the air-conditioning or refrigerating equipment. The disconnecting means is permitted to be installed _____ the air-conditioning or refrigerating equipment, but not on panels that are designed to allow access to the air-conditioning or refrigeration equipment.

   (a) on   (b) within   (c) a or b   (d) none of these

8. Short-circuit and ground-fault protection for an individual motor compressor must not exceed _____ percent of the motor-compressor rated-load current or branch-circuit protection current, whichever is greater.

   (a) 80   (b) 125   (c) 175   (d) 250

9. The rating of the branch-circuit short-circuit and ground-fault protection device for an individual motor-compressor must not exceed _____ percent of the rated-load current or branch-circuit selection current, whichever is greater, if the protection device will carry the starting current of the motor.

   (a) 100   (b) 125   (c) 175   (d) none of these

10. Branch-circuit conductors supplying a single motor-compressor must have an ampacity not less than _____ percent of either the motor-compressor rated-load current or the branch-circuit selection current, whichever is greater.

    (a) 125   (b) 100   (c) 250   (d) 80

11. Conductors supplying more than one motor-compressor must have an ampacity not less than the sum of the rated load or branch-circuit current ratings, whichever is larger, of all the motor- compressors plus the full-load currents of any other motors, plus _____ percent of the highest motor or motor compressor rating in the group.

    (a) 80   (b) 50   (c) 25   (d) 100

12. A hermetic motor-compressor controller must have a _____ current rating not less than the respective nameplate rating(s) on the compressor.

    (a) continuous-duty full-load   (b) locked-rotor   (c) a or b   (d) a and b

13. The rating of the attachment plug and receptacle must not exceed _____ at 250V for a cord-and-plug connected air conditioner.

    (a) 15A   (b) 20A   (c) 30A   (d) 40A

14. The total rating of a cord-and-plug connected room air conditioner, connected to the same branch circuit which supplies lighting units, other appliances, or general use receptacles, must not exceed _____ percent of the branch circuit rating.

    (a) 80   (b) 70   (c) 50   (d) 40

(• Indicates that 75% or fewer exam takers get the question correct)

15. An attachment plug and receptacle is permitted to serve as the disconnecting means for a 1Ø room air conditioner rated 250 volts or less if _____.

   (a) manual controls on the air conditioner are readily accessible within 6 ft of the floor
   (b) an approved operable switch is installed in a readily-accessible location within sight of the air conditioner
   (c) a pushbutton kill switch is installed at the entrance to the room
   (d) a or c

16. When supplying a room air conditioner rated 120 volts, the length of the flexible supply cord must not exceed _____

   (a) 4 ft (b) 6 ft (c) 8 ft (d) 10 ft

Article 445 Generators

Article Overview

This article contains the electrical installation requirements for generators. These requirements include such things as where generators can be installed, nameplate markings, conductor ampacity, and disconnecting means.

Generators are basically motors that operate in reverse—they produce electricity when rotated, instead of rotating when supplied with electricity. Article 430, which covers motors, is the longest article in the *NEC*. Article 445, which covers generators, is one of the shortest. At first, this might not seem to make sense. But you don't need to size and protect conductors to a generator. You do need to size and protect them to a motor.

Generators need overload protection, and it's necessary to size the conductors that come from the generator. But these considerations are much more straightforward than the equivalent considerations for motors. Before you study Article 445, take a moment to read the definition of "Separately Derived System" in Article 100.

Questions

1. Constant-voltage generators must be protected from overloads by _____ or other acceptable overcurrent protective means suitable for the conditions of use.

    (a) inherent design   (b) circuit breakers   (c) fuses   (d) any of these

2. The ampacity of ungrounded (phase) conductors from the generator terminals to the first overcurrent protection devices must not be less than _____ percent of the nameplate rating of the generator.

    (a) 75   (b) 115   (c) 125   (d) 140

3. Unless two restrictive conditions exist, a generator must be equipped with one or more disconnecting means to disconnect the generator, its protective devices, and all control apparatus entirely from the circuits supplied by the generator.

    (a) True   (b) False

(• Indicates that 75% or fewer exam takers get the question correct)

Article 450 Transformers and Transformer Vaults

Article Overview

Article 450 opens by saying, "This article covers the installation of all transformers." Then it lists eight exceptions. So what does Article 450 really cover? Essentially, it covers power transformers, transformer vaults, and most kinds of lighting transformers.

One of the main concerns with transformers is the prevention of overheating. The *NEC* doesn't completely address this issue. Article 90 explains that the *NEC* isn't a design manual and that it assumes the person using the *Code* has a certain level of expertise. Proper transformer selection is an important part of preventing transformer overheating.

The *NEC* assumes you have already selected a transformer suitable to the load characteristics. For the *Code* to tell you how to do that would push it into the realm of a design manual. Article 450 then takes you to the next logical step—providing overcurrent protection and the proper connections. But Article 450 doesn't stop there; 450.9 provides ventilation requirements.

Part I of Article 450 contains the general requirements such as guarding, marking, and accessibility. Part II contains the requirements for different types of transformers, and Part III provides requirements for transformer vaults.

Questions

1. According to Article 450, a transformer rated 600V, nominal, or less, and whose primary current rating is 9A or more, is protected against overcurrent only when _____.

   (a) an individual overcurrent device on the primary side is set at not more than 125 percent of the rated primary current of the transformer.
   (b) a secondary overcurrent device is set at not more than 125 percent of the rated secondary current of the transformer, and a primary overcurrent device is set at not more than 250 percent of the rated primary current of the transformer.
   (c) a or b
   (d) none of these

2. For a transformer rated 600V, nominal, or less, if the primary overcurrent protection device is sized at 250 percent of the primary current, what size secondary overcurrent protection device is required if the secondary current is 42A?

   (a) 40A   (b) 70A   (c) 60A   (d) 90A

3. What size "primary only" overcurrent protection is required for a 600 volt, 45 kVA transformer that has a primary current rating of 54A?

   (a) 70   (b) 80   (c) 90   (d) 100

4. Transformers with ventilating openings must be installed so that the ventilating openings _____.

   (a) are a minimum 18 in. above the floor   (b) are not blocked by walls or obstructions
   (c) are aesthetically located   (d) are vented to the exterior of the building

5. Indoor transformers of greater than _____ rating must be installed in a transformer room of fire-resistant construction.

   (a) 35,000 kVA   (b) 87½ kVA   (c) 112½ kVA   (d) 75 kVA

(• Indicates that 75% or fewer exam takers get the question correct)

6. Each doorway leading into a transformer vault from the building interior must be provided with a tight-fitting door having a minimum fire rating of _____ hours.

   (a) 2　　　　　　　　(b) 4　　　　　　　　(c) 5　　　　　　　　(d) 3

7. Personnel doors for transformer vaults must _____ and be equipped with panic bars, pressure plates, or other devices that are normally latched but open under simple pressure.

   (a) be clearly identified　　(b) swing out　　(c) a and b　　(d) a or b

(• Indicates that 75% or fewer exam takers get the question correct)

Article 460 Capacitors

Article Overview

This article covers the installation of capacitors, including those in hazardous (classified) locations as described by Articles 501 through 503.

Capacitors store energy. Thus, simply disconnecting capacitors doesn't de-energize them. Capacitors have requirements for ampacity, overcurrent protection, disconnecting means, and marking. The requirements for capacitors under 600V are less stringent than for those over 600V.

Questions

1. Capacitors containing more than _____ of flammable liquid must be enclosed in vaults or outdoor fenced enclosures.

    (a) 10 gallons (b) 5 gallons (c) 3 gallons (d) 11 gallons

2. Capacitors must be _____ so that persons cannot come into accidental contact or bring conducting materials into accidental contact with exposed energized parts, terminals, or buses associated with them.

    (a) enclosed (b) located (c) guarded (d) any of these

3. The ampacity of capacitor circuit conductors must not be less than _____ percent of the rated current of the capacitor.

    (a) 100 (b) 115 (c) 125 (d) 135

4. An overcurrent device must be provided in each ungrounded conductor for each capacitor bank. The rating or setting of the overcurrent device must be _____.

    (a) 20A
    (b) as low as practicable
    (c) 400% of conductor ampacity
    (d) 100A

5. A separate overcurrent device is not required for a capacitor connected on the load side of a motor overload protective device.

    (a) True (b) False

6. A disconnect must be provided in each ungrounded conductor for each capacitor bank, and must _____.

    (a) open all ungrounded conductors simultaneously
    (b) be permitted to disconnect the capacitor from the line as a regular operating procedure
    (c) be rated no less than 135 percent of the rated current of the capacitor
    (d) all of these

7. A separate disconnecting means is required where a capacitor is connected on the load side of a motor controller

    (a) True (b) False

8. Where a motor installation includes a capacitor connected on the load side of the motor overload device, the rating or setting of the motor overload device must be based on the improved power factor of the motor circuit.

    (a) True (b) False

(• Indicates that 75% or fewer exam takers get the question correct)

9. Where a motor installation includes a capacitor connected on the load side of the motor overload device, the effect of a capacitor must be disregarded in determining the motor circuit conductor size in accordance with 430.22.

(a) True (b) False

Understanding the National Electric Code, Volume 1
FINAL EXAM

1. A _____ receptacle without GFCI protection can be located in a dwelling unit garage to supply one appliance, which is not easily moved, if the receptacle is located within the dedicated space for the appliance.

   (a) multioutlet (b) duplex (c) single (d) none of these

2. When replacing a nongrounding-type receptacle in a bedroom of a dwelling unit where no grounding means exists in the receptacle enclosure, you must use a _____.

   (a) nongrounding receptacle (b) grounding receptacle (c) GFCI-type receptacle (d) a or c

3. The service disconnecting means must plainly indicate whether it is in the _____ position.

   (a) open or closed (b) tripped (c) up or down (d) correct

4. The largest size grounding electrode conductor required for any service is a _____ copper.

   (a) 6 AWG (b) 1/0 AWG (c) 3/0 AWG (d) 250 kcmil

5. A 100 ft vertical run of 4/0 AWG copper requires the conductors to be supported at _____ locations.

   (a) 4 (b) 5 (c) 2 (d) none of these

6. A receptacle outlet must be installed in dwelling units for every kitchen and dining area countertop space _____, and no point along the wall line can be more than 2 ft, measured horizontally, from a receptacle outlet in that space.

   (a) wider than 10 in. (b) wider than 3 ft (c) 18 in. or wider (d) 12 in. or wider

7. Electric wiring in the air-handling area beneath raised floors for data-processing systems is permitted in accordance with Article 645.

   (a) True (b) False

8. Metal enclosures and raceways for conductors added to existing installations of _____, which do not provide an equipment ground are not required to be grounded if they are less than 25 ft long, they are free from probable contact with grounded conductive material, and are guarded against contact by persons.

   (a) nonmetallic-sheathed cable (b) open wiring (c) knob-and-tube wiring (d) all of these

9. Receptacles installed outdoors, in a location protected from the weather or other damp locations, must be in an enclosure that is _____ when the receptacle is covered.

   (a) raintight (b) weatherproof (c) rainproof (d) weathertight

10. Which rooms in a dwelling unit must have a switch-controlled lighting outlet?

    (a) Every habitable room (b) Bathrooms (c) Hallways and stairways (d) all of these

11. A hermetic motor-compressor controller must have a _____ current rating not less than the respective nameplate rating(s) on the compressor.

    (a) continuous-duty full-load (b) locked-rotor (c) a or b (d) a and b

12. Dwelling unit or mobile home feeder conductors need not be larger than the service conductors and are permitted to be sized according to 310.15(B)(6).

    (a) True (b) False

13. Luminaires are permitted to be installed in a commercial cooking hood where specific conditions are met, including the requirement that the luminaire be identified for use within a _____ cooking hood.

    (a) nonresidential (b) commercial (c) multifamily (d) all of these

14. Service lateral conductors that supply power to limited loads of a single branch circuit must not be smaller than _____.

    (a) 4 AWG copper (b) 8 AWG aluminum (c) 12 AWG copper (d) none of these

15. Surface metal raceway enclosures providing a transition from other wiring methods must have a means for connecting a(n) _____.

    (a) grounded neutral conductor
    (b) ungrounded conductor
    (c) equipment grounding conductor
    (d) all of these

16. The feeder demand load for four 6 kW cooktops is _____ kW.

    (a) 17 (b) 4 (c) 12 (d) 24

17. The minimum clearance for overhead conductors not exceeding 600V that pass over commercial areas subject to truck traffic is _____.

    (a) 10 ft (b) 12 ft (c) 15 ft (d) 18 ft

18. The rating or setting of an overcurrent protection device for a 16.3A single nonmotor-operated appliance must not exceed _____.

    (a) 15A (b) 35A (c) 25A (d) 45A

19. There must be no more than _____ disconnects installed for each electric supply.

    (a) two (b) four (c) six (d) none of these

20. A bare 4 AWG copper conductor installed near the bottom of a concrete foundation or footing that is in direct contact with the earth may be used as a grounding electrode when the conductor is at least _____ in length.

    (a) 25 ft (b) 15 ft (c) 10 ft (d) 20 ft

21. A grounding-type receptacle can replace a nongrounding-type receptacle at an outlet box that does not contain an equipment grounding conductor if the equipment grounding conductor is connected to the _____.

    (a) grounding electrode system as described in 250.50
    (b) grounding electrode conductor
    (c) equipment grounding terminal bar within the enclosure where the branch circuit for the receptacle originates
    (d) any of these

22. A single piece of equipment consisting of a multiple receptacle comprised of _____ or more receptacles must be computed at not less than 90 VA per receptacle.

    (a) 1 (b) 2 (c) 3 (d) 4

23. Connection of conductors to terminal parts must ensure a thoroughly good connection without damaging the conductors and must be made by means of _____.

    (a) solder lugs (b) pressure connectors (c) splices to flexible leads (d) any of these

24. Contact devices or yokes designed and listed as self-grounding are permitted in conjunction with the supporting screws to establish the grounding circuit between the device yoke and flush-type boxes.

    (a) True (b) False

25. Equipment such as raceways, cables, wireways, cabinets, panels, etc. can be located above or below other electrical equipment when the associated equipment does not extend more than _____ from the front of the electrical equipment.

    (a) 3 in. (b) 6 in. (c) 12 in. (d) 30 in.

26. Feeder and service-entrance conductors with demand loads determined by the use of 220.82 are permitted to have the _____ load determined by 220.61.

    (a) feeder (b) circuit (c) neutral (d) none of these

27. Flexible cords approved for and used with a specific listed appliance or portable lamp are considered to be protected when _____.

    (a) not more than 6 ft in length
    (b) 20 AWG and larger
    (c) applied within the listing requirements
    (d) 16 AWG and larger

28. For circuits over 250 volts-to-ground (277/480V), electrical continuity can be maintained between a box or enclosure where no oversized, concentric, or eccentric knockouts are encountered, and a metal conduit by _____.

    (a) threadless fittings for cables with metal sheath
    (b) double locknuts on threaded conduit (one inside and one outside the box or enclosure)
    (c) fittings that have shoulders that seat firmly against the box with a locknut on the inside or listed fittings identified for the purpose
    (d) all of these

29. Surface extensions from a flush-mounted box must be made by mounting and mechanically securing an extension ring over the flush box.

    (a) True (b) False

30. Type AC cable must be supported and secured at intervals not exceeding 4½ ft and the cable must be secured within _____ of every outlet box, cabinet, conduit body, or other armored cable termination.

    (a) 4 in. (b) 8 in. (c) 9 in. (d) 12 in.

31. Type NM cable, installed within accessible ceilings for the connections to luminaires and equipment, does not need to be secured within 12 in. from the luminaire or equipment when the free length does not exceed _____.

    (a) 4½ ft (b) 2½ ft (c) 3½ ft (d) any of these

32. When applying the general provisions for receptacle spacing to the rooms of a dwelling unit, which require receptacles in the wall space, no point along the floor line in any wall space of a dwelling unit may be more than _____ from an outlet.

    (a) 12 ft (b) 10 ft (c) 8 ft (d) 6 ft

33. Where a branch circuit supplies continuous loads, or any combination of continuous and noncontinuous loads, the rating of the overcurrent device must not be less than the noncontinuous load plus 125 percent of the continuous load.

    (a) True (b) False

34. Where the load is computed on volt-amperes per square meter or square foot basis, the wiring system up to and including the branch-circuit _____ must be provided to serve not less than the calculated load.

    (a) wiring (b) protection (c) panelboard(s) (d) all of these

35. A contact device installed at an outlet for the connection of an attachment plug is known as a(n) _____.

    (a) attachment point (b) tap (c) receptacle (d) wall plug

36. A form of general-use switch constructed so that it can be installed in device boxes or on box covers, or otherwise used in conjunction with wiring systems recognized by the *Code* is called a _____ switch.

    (a) transfer (b) motor-circuit (c) general-use snap (d) bypass isolation

37. Alternating current general-use snap switches, suitable only for use on ac circuits, can control _____.

    (a) resistive and inductive loads that do not exceed the ampere and voltage rating of the switch
    (b) tungsten-filament lamp loads that do not exceed the ampere rating of the switch at 120V
    (c) motor loads that do not exceed 80 percent of the ampere and voltage rating of the switch
    (d) all of these

38. Circuit breakers must be marked with their ampere rating in a manner that will be durable and visible after installation. Such marking can be made visible by removal of a _____.

    (a) trim (b) cover (c) box (d) a or b

39. Circuit-protective devices are used to clear a fault without the occurrence of extensive damage to the electrical components of the circuit. Faults can occur between two or more of the _____ or between any circuit conductor and the grounding conductor or enclosing metal raceway.

    (a) bonding jumpers (b) grounding jumpers (c) wiring harnesses (d) circuit conductors

40. Fuseholders of the Edison-base type must be installed only where they are made to accept _____ fuses by the use of adapters.

    (a) Edison-base (b) medium-base (c) heavy-duty base (d) Type S

41. Lighting track must not be installed less than _____ above the finished floor except where protected from physical damage or track operating at less than 30 volts rms, open-circuit voltage.

    (a) 4 ft (b) 5 ft (c) 5½ ft (d) 6 ft

42. Materials such as straps, bolts, screws, etc. that are associated with the installation of IMC in wet locations are required to be _____.

    (a) weatherproof (b) weathertight (c) corrosion-resistant (d) none of these

43. One of the requirements that permit conductors supplying a transformer to be tapped, without overcurrent protection at the tap, is that the conductors supplied by the _____ of a transformer must have an ampacity, when multiplied by the ratio of the primary-to-secondary voltage, of at least one-third the rating of the overcurrent device protecting the feeder conductors.

    (a) primary (b) secondary (c) tertiary (d) none of these

44. Premises wiring must not be electrically connected to a supply system unless the supply system contains, for any grounded neutral conductor of the interior system, a corresponding conductor that is ungrounded.

    (a) True (b) False

45. The NEC is _____.

    (a) intended to be a design manual
    (b) meant to be used as an instruction guide for untrained persons
    (c) for the practical safeguarding of persons and property
    (d) published by the Bureau of Standards

46. The use of NUCC is permitted _____.

    (a) for direct-burial underground installations
    (b) to be encased or embedded in concrete
    (c) in cinder fill
    (d) all of these

47. To prevent water from entering service equipment, service-entrance conductors must _____.

    (a) be connected to service-drop conductors below the level of the service head
    (b) have drip loops formed on the service-entrance conductors
    (c) a or b
    (d) a and b

48. Where an ac system operating at less than 1,000V is grounded at any point, the _____ conductors must be run to each service disconnecting means and must be bonded to each disconnect enclosure.

    (a) ungrounded (b) grounded (c) grounding (d) none of these

49. Where nails or screws are likely to penetrate nonmetallic-sheathed cable or electrical nonmetallic tubing installed through metal framing members, a steel sleeve, steel plate, or steel clip not less than _____ in thickness must be used to protect the cable or tubing. A thinner plate that provides equal or better protection may be used if listed and marked.

    (a) 1/16 in. (b) 1/8 in. (c) 1/2 in. (d) none of these

50. A circuit breaker is a device designed to _____ a circuit by nonautomatic means and to open the circuit automatically on a predetermined overcurrent without damage to itself when properly applied within its rating.

    (a) blow (b) disconnect (c) connect (d) open and close

51. A receptacle outlet for the laundry is not required in a dwelling unit in a multifamily building when laundry facilities that are available to all building occupants are provided on the premises.

    (a) True (b) False

52. Appliances that have a unit switch with a marked _____ setting that disconnects all the ungrounded conductors is permitted to serve as the disconnecting means for the appliance, where other means of disconnection are also provided in accordance with 422.34.

    (a) "on" (b) "off" (c) "on/off" (d) all of these

53. Capable of being reached quickly for operation, renewal, or inspections without resorting to portable ladders and such is known as _____.

    (a) accessible (equipment)
    (b) accessible (wiring methods)
    (c) accessible, readily
    (d) all of these

54. Grounding and bonding conductors cannot be connected by _____.

    (a) pressure connections (b) solder (c) lugs (d) approved clamps

55. In electrical nonmetallic tubing, the maximum number of bends between pull points cannot exceed _____ degrees, including any offsets.

    (a) 320 (b) 270 (c) 360 (d) unlimited

56. Metal or nonmetallic poles over 20 ft in height above grade that support luminaires must meet which of the following requirements? _____.

    (a) They must have an accessible handhole (sized 2 x 4 in.) with a raintight cover
    (b) The grounding terminal must be accessible from the handhole
    (c) a and b
    (d) none of these

57. One of the requirements that must be met to use steel or aluminum cable tray systems as equipment grounding conductors, is that the cable tray sections and fittings have been _____ marked to show the cross-sectional area of metal in channel cable trays, or cable trays of one-piece construction and total cross sectional area of both side rails for ladder or trough cable trays.

    (a) legibly (b) durably (c) a or b (d) a and b

58. Torque requirements for motor control circuit device terminals must be a minimum of _____ lbs-inch (unless otherwise identified) for screw-type pressure terminals used for 14 AWG and smaller copper conductors.

    (a) 7 (b) 9 (c) 10 (d) 15

59. Type MC cable must be supported and secured at intervals not exceeding _____.

    (a) 3 ft (b) 6 ft (c) 4 ft (d) 2 ft

60. UF cable used with a 24V landscape lighting system is permitted to have a minimum cover of _____.

    (a) 6 in. (b) 12 in. (c) 18 in. (d) 24 in.

61. When multiple ground rods are used for a grounding electrode, they must be separated not less than _____ apart.

    (a) 6 ft (b) 8 ft (c) 20 ft (d) 12 ft

62. A _____ is a single unit that provides independent living facilities for persons, including permanent provisions for living, sleeping, cooking, and sanitation.

    (a) two-family dwelling (b) one-family dwelling (c) dwelling unit (d) multifamily dwelling

63. A box or conduit body is not required where cables enter or exit from conduit or tubing that is used to provide cable support or protection against physical damage. A fitting must be provided on the end(s) of the conduit or tubing to _____.

    (a) allow for the future connection of a box
    (b) be used for a future pull point
    (c) protect the cable from abrasion
    (d) allow the coupling of another section of conduit

64. A horsepower-rated inverse-time circuit breaker can serve as both a motor controller and disconnecting means if _____.

   (a) it opens all ungrounded conductors
   (b) it is protected by an overcurrent device in each ungrounded conductor
   (c) it is manually operable, or both power and manually operable
   (d) all of these

65. A recessed luminaire (fixture) that is not identified for contact with insulation must have all recessed parts spaced not less than _____ from combustible materials, except for points of support and the trim finishing off the opening in the ceiling or wall.

   (a) ¼ in.   (b) ½ in.   (c) 1¼ in.   (d) 6 in.

66. A strut-type channel raceway cannot be installed _____.

   (a) in concealed locations
   (b) where subject to corrosive vapors if protected solely by enamel
   (c) a or b
   (d) none of these

67. A TVSS device must be listed.

   (a) True   (b) False

68. Bends in flexible metal conduit must _____ between pull points.

   (a) not be made
   (c) be limited to not more than 360 degrees
   (b) not be limited in degrees
   (d) be limited to 180 degrees

69. Bends in LFNC must be made so that the conduit will not be damaged and the internal diameter of the conduit will not be effectively reduced. Bends are permitted to be made only _____.

   (a) manually without auxiliary equipment
   (c) with any kind of conduit bending tool that will work
   (b) with bending equipment identified for the purpose
   (d) by use of an open flame torch

70. Bends in rigid nonmetallic conduit must _____ between pull points.

   (a) not be made
   (c) be limited to not more than 360 degrees
   (b) not be limited in degrees
   (d) be limited to 180 degrees

71. Bends made in USE and SE cable must be made so that the cable will not be damaged. The radius of the curve of the inner edge of any bend during or after installation must not be less than _____ the diameter of the cable.

   (a) 5 times   (b) 7 times   (c) 10 times   (d) 125% of

72. Each current-carrying conductor of a paralleled set of conductors must be counted as a current-carrying conductor for the purpose of applying the adjustment factors of 310.15(B)(2)(a).

   (a) True   (b) False

73. Electrical systems are grounded to the _____ to stabilize the system voltage.

   (a) ground   (b) earth   (c) electrical supply source   (d) none of these

74. Exposed vertical risers of RMC for industrial machinery or fixed equipment can be supported at intervals not exceeding _____ if the conduit is made up with threaded couplings, firmly supported at the top and bottom of the riser, and no other means of support is available.

    (a) 6 ft (b) 10 ft (c) 20 ft (d) none of these

75. HDPE is not permitted to be installed _____.

    (a) where exposed
    (b) within a building
    (c) for conductors operating at a temperature above the rating of the raceway
    (d) all of these

76. Horizontal runs of EMT supported by openings through framing members at intervals not greater than _____, and securely fastened within 3 ft of termination points, are permitted.

    (a) 1.4 ft (b) 12 in. (c) 4 ½ ft (d) 10 ft

77. Insulated conductors and cables exposed to the direct rays of the sun must be _____.

    (a) covered with insulating material that is listed or listed and marked sunlight resistant
    (b) listed and marked sunlight resistant
    (c) listed for sunlight resistance
    (d) any of these

78. Liquidtight flexible metal conduit (LFMC) up to trade size ½ can be used as the equipment grounding conductor if the length in any ground return path does not exceed 6 ft and the circuit conductors contained in the conduit are protected by overcurrent devices rated at _____ or less when the conduit is not installed for flexibility.

    (a) 15A (b) 20A (c) 30A (d) 60A

79. Liquidtight flexible metal conduit is not required to be fastened when used for tap conductors to luminaires up to _____ in length.

    (a) 4½ ft (b) 18 in. (c) 6 ft (d) no limit on length

80. Metal raceways, boxes, fittings, supports, and support hardware can be installed in concrete or in direct contact with the earth or other areas subject to severe corrosive influences, where _____ approved for the conditions, or where provided with corrosion protection approved for the purpose.

    (a) the soil is (b) made of material (c) the qualified installer is (d) none of these

81. Nonmetallic cable trays are permitted in corrosive areas and in areas requiring voltage isolation.

    (a) True (b) False

82. Outlet boxes used at luminaire or lamp holder outlets must be _____.

    (a) designed for the purpose
    (b) metal only
    (c) plastic only
    (d) mounted using bar hangers only

83. Plaster, drywall, or plasterboard surfaces that are broken or incomplete must be repaired so there will be no gaps or open spaces greater than _____ at the edge of a cabinet or cutout box employing a flush-type cover.

    (a) ¼ in. (b) ½ in. (c) ⅛ in. (d) 1/16 in.

84. Single-throw knife switches must be installed so that gravity will tend to close the switch.

    (a) True   (b) False

85. Switchboards, panelboards, industrial control panels, meter socket enclosures, and motor control centers in commercial and industrial occupancies that are likely to require _____ while energized must be field marked to warn qualified persons of the danger associated with an arc flash from line-to-line or ground faults.

    (a) examination
    (b) adjustment
    (c) servicing or maintenance
    (d) a, b, or c

86. Table 430.91 provides the basis for selecting enclosures for use in specific locations other than _____.

    (a) agricultural buildings
    (b) hazardous locations
    (c) recreational vehicle parks
    (d) assembly occupancies

87. The bonding jumper used to bond the metal water piping system to the service must be sized in accordance with _____.

    (a) Table 250.66   (b) Table 250.122   (c) Table 310.16   (d) Table 310.15(B)(6)

88. The derating factors in 310.15(B)(2)(a) apply to a nonmetallic wireway.

    (a) True   (b) False

89. The grounding electrode conductor is the conductor used to connect the grounding electrode to the equipment grounding conductor and the grounded neutral conductor at _____.

    (a) the service
    (b) each building or structure supplied by feeder(s)
    (c) the source of a separately derived system
    (d) all of these

90. The number of fixture wires in a single conduit or tubing must not exceed that permitted by the percentage fill specified in _____.

    (a) Table 1, Chapter 9   (b) Table 250.66   (c) Table 310.16   (d) 240.6

91. The rating of a branch circuit is determined by the rating of the _____.

    (a) ampacity of the largest device connected to the circuit
    (b) average of the ampacity of all devices
    (c) branch-circuit overcurrent protection
    (d) ampacity of the branch circuit conductors according to Table 310.16

92. The term "luminaire" includes "fixture(s)" and "lighting fixture(s)."

    (a) True   (b) False

93. Transformers with ventilating openings must be installed so that the ventilating openings _____.

    (a) are a minimum 18 in. above the floor
    (b) are not blocked by walls or obstructions
    (c) are aesthetically located
    (d) are vented to the exterior of the building

94. Underground raceways and cable assemblies entering a handhole enclosure must extend into the enclosure, but they are not required to be _____.

    (a) bonded
    (b) insulated
    (c) mechanically connected to the handhole enclosure
    (d) below minimum cover requirements after leaving the handhole

95. Unused openings for circuit breakers and switches in switchboards and panelboards must be closed using _____ or other approved means that provide protection substantially equivalent to the wall of the enclosure.

    (a) duct seal and tape　　(b) identified closures　　(c) exothermic welding　　(d) sheet metal

96. Utilities include entities that install, operate, and maintain _____.

    (a) communications systems (telephone, CATV, Internet, satellite, or data services)
    (b) electric supply systems (generation, transmission, or distribution systems)
    (c) Local Area Network wiring on premises
    (d) a or b

97. When counting the number of conductors in a box, a conductor running through the box with an unbroken loop not less than twice the minimum length required for free conductors in 300.14 is counted as _____ conductor(s).

    (a) one　　(b) two　　(c) zero　　(d) none of these

98. When installed in _____, nonmetallic-sheathed cable must be protected from physical damage where necessary by RMC, IMC, Schedule 80 rigid nonmetallic conduit, EMT, guard strips, or other means.

    (a) hazardous locations of commercial garages　　(b) exposed work
    (c) service entrance applications　　(d) motion picture studios

99. Where a grounded neutral conductor is installed and the neutral-to-case bond is not at the source of the separately derived system, the grounded neutral conductor must be routed with the derived phase conductors and must not be smaller than the required grounding electrode conductor specified in Table 250.66, but must not be required to be larger than the largest ungrounded derived phase conductor.

    (a) True　　(b) False

100. Where a motor installation includes a capacitor connected on the load side of the motor overload device, the effect of a capacitor must be disregarded in determining the motor circuit conductor size in accordance with 430.22.

    (a) True　　(b) False

Notes

Notes

Notes

Notes

Notes

Notes